THE STRANGEST ANIMALS IN THE WORLD

Professor S. Talalaj & J. Talalaj

Hill of Content
Melbourne

First published in Australia 1993
by Hill of Content Publishing Co Pty Ltd
86 Bourke Street Melbourne 3000 Australia

Cover photograph: Marine Hatchetfish
Typeset by Advertype Group Pty Ltd
Printed by Griffin Paperbacks

National Library of Australia
Cataloguing-in-Publication data

Talalaj, S. (Stanislaw), 1923-
The strangest animals in the world.

Bibliography.
Includes index.
ISBN 0 85572 231 2.

1. Animals. 2. Animals – Folklore. I. Talal

591

Acknowledgments of Illustrations
The publishers and authors would like to thank the following for permission to use the specified photographs
(*see central photographic section*):

Thorny Devil	(Kathie Atkinson)
Frilled Lizard	(Kathie Atkinson)
Satin Bower Bird	(Australian Government Publishing Service, Bush Dwellers of Australia, Canberra, 1986. Commonwealth of Australia copyright reproduced by permission)

Preface

The writing of this book has been inspired by our unforgettable experiences while watching animals in the wild, particularly in the National Parks of Kenya. This book is not only about the magnificent animals which are well known but particularly about the truly incredible creatures which are unknown to most people. The external features or behaviour of these creatures is so incredible that it is almost impossible to believe that the facts provided are true. The reader will learn, for instance about species whose female inserts a penis-like organ into the male to receive his pocket of sperm. We also describe a mammal which can copulate 157 times in just 55 hours and another species of mammal which is said to copulate non-stop for up to 60 minutes.

We also describe little known facts about the intelligence of certain animals. For instance, we include an animal which is its own doctor. Curing itself by eating a particular plant which it chooses in the forest depending on the ailment. Some of the plants used by this animal were not even known to the local people for their medicinal properties until its behaviour was studied in detail.

We have also outlined some of the amazing abilities of certain animals. Some creatures can for example function with their head cut off for up to seven days.

We explain various secrets and mysteries associated with some members of the animal kingdom. For instance, the behaviour of a certain creature was a secret of international significance kept undisclosed for almost 3,000 years.

While researching this book we have made a fascinating discovery. Some animals which we all greatly admire and love and also those which people despise, have greatly stimulated people's imagination world wide. As a result a truly fascinating set of beliefs and customs associated with religion, magic and folk medicine has evolved. The inclusion of these stories has greatly modified the character of the book which has consequently become a book full of adventure, suspense and surprising and amusing facts. For instance in some regions of the world men deliberately mutilate their sexual organs in order for them to resemble those of certain 'beasts' particularly sexually potent.

Hence it is not just a book dedicated to the lovers of animals but equally to the lovers of history, folklore and anthropology. It will be equally valuable to the dedicated addicts of arm-chair travel.

In closing we would like to thank the countless people working in national parks and zoos world wide who helped us and other people to enjoy the unique beauty of our animal kingdom. We would also like to thank Mrs Daniela Talalaj for her help and useful suggestions during preparation of this book. We would also like to thank our publisher Michelle Anderson for her efforts in making this book more attractive.

Contents

I.

Animals For Which Humans Were Sacrificed

Snakes with Human Wives

The cult of Dangh-Gbe, the sacred python, was once so strong in Benin (formerly Dahomey) in West Africa, that temples were specially built in his honour. To ensure a good harvest, the snake-god was provided with many human 'wives' who were chosen from among the most attractive girls aged between 8 and 12. The girls had to dance at a special ceremony begging the snake-god for rain. After the ceremony, human sacrifices were made to the powerful python-god.

Becoming a wife of the sacred python was a great privilege so devoted parents used to leave their young daughters on the street to attract the attention of the special priestesses responsible for choosing the python-god's wives. These 'wives' were initially involved in highly immoral sexual rites with the priests of the python temples and later they became prostitutes serving the male worshippers of the python god. The children born from these sexual relations were treated as belonging to the snake god.

Pythons were so venerated in some parts of West Africa that the first treaty between the British and the local Nigerian kings had to include a specific clause to ensure the protection of various pythons. Killing pythons, even by accident, was considered a great offence and invited severe punishment. Deliberately killing a python was even punished by death. 'The offender was placed in the hole under a hut . . . thatched with grass that had been generously greased with palm oil, and the hut was set aflame.'

Even in the early nineteenth century, a traveller exploring West Africa witnessed python worship. He saw a number of small temples in which many pythons were living. The pythons were regarded as sacred and were fed with rats and mice by special priests. The temples were visited regularly by the natives, especially the sick and crippled, who tried to reach the sacred snake temple to be cured. Pythons were treated in West Africa with such great respect that a man would have to give way to a python if they met, and the snake was always greeted with the saying 'my father.'

The first Europeans were often unaware of the peculiar snake cult existing in some parts of Africa. The authors were once told in Ghana of a European who accidentally killed a python when driving through the forest. He took the dead snake home to show it to his family. His servant, on seeing the dead snake, panicked and shouted that his master had killed a god. He only calmed down when he learned that it had been an accident. The servant explained that if one intentionally killed a sacred python a calamity to one's family would result and he told them the following story.

Long ago his tribe had been chased by hostile neighbours and while escaping they came to a deep river which was impossible to cross. Suddenly, to their astonishment, the river became so full of pythons that they could walk over the bodies as if on a bridge, and as a result they easily escaped from their oppressors.

Even more fascinating is the story told in the Ashanti district of Ghana. According to this story, the first man and woman on earth had no knowledge of conception so they could not have children. It was a wise python who 'taught them how to procreate by making the couple stand facing each other and spray their bellies with water, then telling them to go home and sleep together.' Following the python's advice, the story says, the woman soon conceived and a beautiful child was born. For this reason pythons are regarded as sacred in this region of West Africa.

Although pythons do not possess venomous fangs they are dangerous because they hug their victims to death. Large animals have been killed by pythons. There was even a report that a large African python overpowered a leopard and swallowed it. Giant pythons are also dangerous to humans. A story is told of a man who was hunting in the jungle in Burma and took refuge under a tree during heavy rain. The rest of the group of hunters tried to find him after the storm was over, but he seemed to have disappeared. Finally they spotted his hat and shoes beneath a tree. Beside them lay a huge python. The body of the missing man was found in the snake's body. The huge python was the reticulated python (*Python reticulatus*) and was some 9 m long.

Pythons are, in fact, among the world's largest snakes and the largest specimen of reticulated python was exactly 10 m long. In Uganda, in 1951, a 13-year-old boy was overpowered and reportedly swallowed by a rock python (*Python sebae*).

There is an interesting python in West Africa called the calabar ground python (*Calabaria reinhardtii*). Its behaviour is unique. When this snake moves it keeps its small head down and its wide, blunt tail up so that the tail resembles its head. In this way the clever snake fools its predators, protecting itself from being attacked from the front. Africans call it the two-headed snake. It spends much of its time underground catching small rodents.

Pythons, and all other snakes, have two penises known as

hemipenes. Each hemipenis is a kind of sac which turns inside out like a glove finger during copulation. The male uses only one of his hemipenes, depending on which side of him the female is on. In order to sexually stimulate the female, the male python uses the rudiments of his pelvic hind limbs to scratch the female's cloacal region.

Female pythons are the only snakes reported to care for their eggs with any sophistication. In a number of python species the female coils herself around the clutch of eggs and remains with them until they hatch. The Indian python (*Python molurus*) regulates her body heat while incubating her eggs. When the temperature outside drops below 30°C she contracts her body muscles at the rate of up to 30 contractions a minute and increases the temperature by up to 7°C. This is the only species of reptile that incubates its eggs in this particular way.

Bizarre Sacrifice

The jaguar (*Panthera onca*), the New World's largest cat, played a significant role in the rituals of ancient people. The Aztecs' most important god, the Sun god Tezcatlipoca, was believed to appear in the form of a powerful jaguar. Aztecs believed that this god could destroy the whole world if he wished to and they feared him greatly. To please this powerful god, every year the Aztecs offered him a human on a sacrificial altar. The most handsome young prisoner was always chosen as the sacrificial victim. The extraordinary feature of this amazing custom was that before being sacrificed, the man was supposed to personify the god for whom he was to be killed. For a whole year the chosen man was treated like a god. He was dressed in truly extravagant and beautiful clothes and his body was decorated with flowers. He had eight servants ready to fulfil his desires.

Then, twenty days before the fatal day, the young human-god was given four beautiful girls who became his wives. These women became goddess-wives of the Sun god Tezcatlipoca during this short marriage. Finally, after the whole year had passed, the man was led to the temple in a splendid procession and placed on a sacrificial altar. With a sharp knife the priest cut off his breast and tore out his still palpitating heart, which was ceremoniously offered to the Sun god, Tezcatlipoca.

The veneration of the jaguar developed among the ancient Maya to such an extent that people tried everything to look like the sacred beast. Parents deformed the heads of their infants so that they would become flattened like those of jaguars. This was accomplished by using special boards which were bound around the baby's skull, to give it this 'much-desired' shape. In Chichen Itza, one of the greatest cities of the Maya, a great pyramid temple was devoted to the jaguar. The interior of the temple was richly decorated with marching jaguars and in the throne room there was a red life-size effigy of a jaguar. This spectacular figure had jade eyes and a jade-encrusted tail.

The strange jaguar cult survived the Spaniards' conquest, although it was severely persecuted and was performed in secret for several centuries until it almost vanished. In some regions of Mexico it has survived in the form of an innocent, traditional dance known as the 'jaguar dance' and in many regions of South America the jaguar is still believed to possess supernatural powers. No wonder that when an Indian makes his arrow poison he adds a few jaguar hairs to strengthen its killing power.

In some regions, Indians still believe that when a local shaman dies he miraculously changes into a jaguar that hates people and attacks them at night. If a man kills a jaguar it is customary for him to take the dead beast's name and perform certain rituals in order to transform himself into a jaguar. The man has to live for a specified period of time in a jaguar temple where heads of the animal are kept at the altars.

The jaguar is regarded in many regions of South America as a symbol of strength and jaguar teeth are commonly used by many Indians to adorn their bodies. Some naked Amazonian Indians never forget to wear beads of jaguar teeth round their necks as charms.

Although jaguars were hunted by Indians, the communal feasts were always held outside the villages. It was believed that if the meat of the jaguar was eaten in the village it would offend the animal's soul and cause its revenge. The killing of a jaguar was such a great event that it was followed by joyful feasts, which often lasted several days.

The jaguar, which can measure some 2.4 m in length, is remarkably strong. It can kill a horse, pull it through the forest, tow the body across a river and then start to devour it in a chosen safe place. Jaguars are excellent swimmers and they spend much of their time in rivers and ponds where they catch fish with their paws. There is a strong belief among the Amazonian Indians that a jaguar uses its tail to catch fish. This may be true because some reports indicate that jaguars 'deliberately' twitch their tails in the water to lure fish and then strike them with their paws. The jaguar is also known to dig up and eat the eggs of turtles on the beaches of Central America.

Shark Kisser

No other sea animal frightens people as much as the shark. In many islands of the Pacific the shark was regarded as a powerful and vengeful god to be appeased at all costs. It was insufficient to worship the evil shark-god — a human sacrifice was also necessary.

On one Pacific island a special sacrificial ceremony was organised by high priests. The person to be sacrificed was caught by a noose resembling a shark snare. The noose was hurled at the crowd of people who had gathered to witness the cruel event. The overpowered victim was strangled to death and cut into pieces, which were then thrown into the sea to please the shark-god.

Humans also used to be offered to the shark-god in Hawaii. When the US Navy started to construct its base at Pearl Harbor, the remains of an old shark pen were discovered. It was built of rocks arranged in a circle with an opening at one end. It was discovered that, in the past, Hawaiian kings used this pen as a sacrificial altar to the shark-god.

Hawaiian warriors used to participate in the most dangerous contests imaginable. They had to fight a deadly shark using a short, wooden dagger which had a shark's tooth on the tip. It was a life-or-death contest. The fighting man had to swim towards the large shark and, before the monster fish could attack, the warrior had to dive beneath the fish and stab the dagger into its belly. Such a gladiator had very little chance of escaping unhurt and there was a common belief in those days that this famous contest was just another way of making another sacrifice to the demanding shark-god.

There used to be a strange belief in Fiji that the most deadly shark could be overpowered by using special charms. The shark charmers specialised in a ritual known as shark kissing. The ritual was witnessed once in the 1930s by A. L. Laplante, a Catholic priest, who reported: 'The natives drive the sharks into a large net, the shark kissers wade out, seize the man-eaters, kiss them on their up-turned bellies and fling them on their back. I don't know how they do it, but among the natives, it is taken for granted that once a shark is kissed — upside down — that is the end of it.' It was believed that after such a ceremony the sea was safe for the fishermen.

Even religious wars were fought in the Pacific due to the shark-gods. Such wars erupted when a member of one tribe defied the shark taboo of another by catching one of their sacred sharks, so committing a sacrilege. As a result a delegation would be sent to the tribe whose member was responsible for the killing. If the guilty tribe refused to stop this desecration a holy war would be declared.

Sharks were blamed in the past for high losses of sailors, especially during World War II. In July 1945 the US ship *Indianapolis* was sunk by a Japanese submarine in the Pacific. Only 316 men were rescued while about a hundred men were taken by hungry sharks, despite the fact that every US sailor was provided with a special shark repellent called a 'shark chaser'. This substance was obtained from rotting shark flesh and mixed with a black dye. The shark repellent was kept in special packages attached to the sailors' life-jackets. It was assumed that when the powder was thrown into the water a huge cloud would hide the man from an attacking fish. It was revealed after the war, however, that this 'repellent' did not work at all and that it was simply a bluff to boost the sailors' morale.

An interesting discovery has been made in recent years. It has been found that certain fish, Moses sole (*Pardachirus mormoratus*), have the power to deter deadly sharks. This small fish, which lives in the Red Sea, was investigated in the marine laboratory at Elat (Israel). When one of these fish was placed in a pool containing a shark, it was not attacked. It was found that the small sole discharges a milky substance into the water which has the remarkable effect of paralysing the shark's jaws so that it is unable to devour its easy prey. Japanese researchers claimed in 1986 that they had synthesised a substance based on the sole fish deterrent. How effective it would be as a practical shark deterrent remains to be seen.

Although sharks have a very bad reputation and are considered to be the most dangerous of sea creatures, only some 27 out of 250 known shark species are harmful to man. In the whole history of Australia sharks are reported to have killed 150 people. The most feared is the great white shark (*Carcharodon carcharias*) called the white death. It can reach 8 m in length and can weigh over a tonne. This shark is a man-attacker and is the nightmare of swimmers and divers.

The man-attacking habits of certain sharks have been put to man's use. In the early 19th century, the governor of one of the main prisons in Tasmania became annoyed about the frequency with which prisoners escaped by swimming across the heavily-guarded, narrow channel. He ordered all kitchen refuse and garbage to be dumped into the sea around the prison walls. No-one entered the water again, as it has been 'watched' ever since by hordes of hungry sharks.

The shark's taste for human flesh was utilised in the past by an unscrupulous Dutchman in Nicaragua. When he learned that the local Indians living around the shores of Lake Nicaragua had a habit of throwing their dead into the lake to be devoured by sharks, he decided to make a fortune. He noticed that the corpses were richly decorated with jewellery so he decided to extract the 'wealth' from the sharks' stomachs. This 'fishing' business brought him a fortune, but his practice was finally discovered by the Indians. His house was destroyed and the greedy Dutchman, the story says, was killed by avenging Indians.

Some sharks are extremely large. The largest is the whale shark (*Rhincodon typus*) which can reach 15 m in length and weigh 15 tonnes. Strangely, this shark only feeds on plankton and is so harmless that it even allows swimmers to climb and travel on its huge body. In order to obtain sufficient tiny plankton to feed itself, the whale shark filters an enormous quantity of water.It can filter some 400,000 gallons of water in just one hour. This world's largest fish species also produces the world's largest eggs which can be up to 35 cm long.

Some dangerous sharks are a very peculiar shape. Perhaps the strangest among them are the hammerhead sharks which greatly differ in appearance from other sharks. They include five species which have heads of an unusual shape. The head is so broad that viewed from above it resembles a large hammer, hence its common name. The eye arrangement is also unique as they are located at the extremities of these lateral head extensions. Some hammerhead sharks, such as the smooth hammerhead (*Sphyrna zygaena*), may reach a length of some 4 m and its 'hammer' is so wide that its eyes are some 100 cm apart. Biologists claim that the hammerhead shark's unusual eye position gives it a much wider view and it can detect its prey or predator more easily. Hammerheads have such sharp teeth that Maoris use them as razor blades. These strange sharks are voracious and have been known to attack man.

Sharks have many remarkable features which puzzle biologists. They have a fantastic ability to detect traces of human blood in salt water: they respond to a dilution of one part human blood to a million parts water. This is why a wounded man in the sea has such a low chance of escaping shark attack. Some sharks have an enormous bite. A 2 m long shark can bite a man in half. Some sharks' teeth are so sharp that they can cut through the thick shell of a huge turtle. Dentists would be surprised by the incredible number of teeth some shark species use and replace. For example, a tiger shark may produce and shed 24,000 teeth during its ten year life span.

Many shark species are known for their incredible appetites and ability to devour and swallow almost any object they meet in the sea. The study of the stomach contents of some sharks has revealed that they have swallowed such unexpected items as a lady's overcoat, a car licence and even a dog with its collar. Their urge for flesh is so strong that the first- born baby shark may devour its brothers and sisters.

Sharks are quite intelligent creatures. Some sharks kept in captivity learnt to press levers to obtain food. Their learning ability could be compared to that of a laboratory rat.

It should be noted that during the Second World War most of the important supplies of vitamin A were obtained from the livers of a certain shark species. The Japanese utilise shark liver for the isolation of medicinally active ingredients. A substance called squalene is made into capsules called Marine Gold which are widely sold in Japan as a panacea for all ills, including cancer. Biological study of sharks has revealed that they are immune to most diseases and cancer is extremely rare among sharks. Scientists believe that an anticancer active compound could be isolated from shark blood. If this were successful a 'horrible man-eater' could become man's best friend and life saver.

Infant Sacrifice

Crocodiles, which are among the most ruthless killers responsible for the death of thousands of people, were once the object of veneration in some cultures. In fact special temples were erected in ancient Egypt for the worship of the crocodile god Sobek. In the ancient Egyptian city Arsinoe (called Crocodilopolis by the Greeks) there was a crocodile which was believed to be the incarnated god Sobek. This reptile wore golden rings beset with precious jewels and golden bracelets adorned his forelegs. He was kept in a special lake and priests fed him daily. According to Herodotus (484–625 BC): 'Priests approached him and while one of them held open its jaws, another put in the cake and the meat and poured the honey wine. After that the animal dived into the lake and swam towards the opposite shore. Another visitor arrived, also bringing his offerings. The priest ran round the lake with the food he had brought and fed it to the crocodile in the same manner.'

When a sacred crocodile died it was mummified, like a human aristocrat, and placed in a specially built tomb. The crocodile cult must have been very strong since thousands of crocodile mummies have been excavated, including the mummies of new-born crocodiles. Even the eggs of the sacred animals were found in some crocodile tombs.

In some regions of the world, crocodile worship developed to such an extent that humans were sacrificed to please the sacred animals. When Chepelier, the French naturalist and explorer, visited Madagascar in the early nineteenth century he was astonished to learn that humans were offered to a sacred croc. A chief in the Bay of Antogil had a sacred crocodile in his pool and believed this animal to be his ancestor. Each year the chief ordered his people to perform a cruel sacrifice in which a young man and woman were offered to the ferocious reptile. The chosen victims were adorned with precious jewellery and then suspended over the pool. People watched with excitement while the sacred crocodile accepted the offering. Also, in some regions of India where a crocodile was worshipped by some Hindu sects, mothers used to give their own infants as sacrifices to the demanding crocodile god.

In many parts of Madagascar crocodiles were respected and considered to be a symbol of wisdom and good judgement. It was believed that crocodiles knew how to distinguish between right and wrong and that they only attacked those who were guilty of crimes. There even used to be a peculiar form of judgement called 'ordeal by crocodile'. People were so convinced that the crocodile was the best judge that those accused of a crime voluntarily submitted themselves to the strange ordeal. If the person accused of a crime was not attacked and mauled by the beast, he was considered not guilty. Most people, however, were found guilty by the hungry 'judge'.

Fear of this reptile has been exploited by some people for their own purpose. There was a famous witch-doctor in a small village near Tamale in northern Ghana who kept a croc to cure his patients. It was a most peculiar treatment. While the doctor talked to a patient, especially one who suffered mental problems, the door of his 'clinic' would suddenly open and a big crocodile would appear ready to catch the victim. This was a prearranged 'shock' therapy and the patient was of course 'rescued' by the doctor. It is said that after this shock of sudden, imminent death and miraculous survival, the patient usually returned home completely healthy. Ironically, before the authors left Ghana in 1966, the famous healer was devoured by his 'befriended' croc.

Perhaps the most horrible ordeal involving ferocious crocodiles took place during the Second World War in Burma in 1944. About a thousand Japanese soldiers were forced by the British to retreat to mangrove swamps inhabited by crocodiles. Most of them perished during the night when attacked by the hungry beasts. It was a real 'hell on earth' and as one report indicated: 'The men of the British ships were horrified by the din that went on through the night, the screams and yells of the soldiers, the thrashing of the crocodiles as they rolled over, tearing their victims apart. When morning came there were only twenty survivors.'

The most dangerous crocodile is the saltwater or estuarine crocodile (*Crocodylus porosus*) which inhabits swamps, rivers and estuaries around coastal areas from Northern Australia to India. It can attain a remarkable length of up to 8 metres and is the world's largest reptile. It is the most feared animal as it is the greatest man-eater of all crocodiles. They are known to unexpectedly attack people and are so powerful that they can pull a horse from the estuarine bank and eat it. This dangerous beast is responsible for several human deaths in Australia and public response was such that so many crocodiles were killed they faced extinction in this country. To prevent this, the saltwater crocodile became a protected animal in Australia in 1972.Now these dangerous crocs are so common in the Northern Territory that special warning signs to visitors have been placed in many areas. Many people believe that crocs only attack from the water. This is untrue since they are known to 'wander' far away from water and attack in the jungle where no one expects them to be.

Biologists who study these 'ancient' reptiles are surprised by some of their habits. The female of the Nile crocodile has interesting breeding habits. She deposits her eggs in deep holes which she carefully covers with soil. When the youngsters are about to hatch they emit characteristic squeals as if calling their parents to help them. The female rushes to the nest and helps them to emerge. She then takes them one by one in her jaw and carries them to water where they are 'spat' to freedom. Crocodile hunters, unaware of the maternal care of these crocs believed that crocodile mothers had a habit of eating their babies.

In some crocodile species even the father helps in the final stage of hatching. Before the eggs hatch he may take one into his huge mouth and gently crack its shell to facilitate the birth. An unusual task for powerful jaws which crush with ease the bones of huge beasts.

Incredible Beast

The tiger (*Panthera tigris*), the giant cat of Asia, has frightened indigenous people since the earliest times. The Golds, Orotchons and Tungus of eastern Siberia and Manchuria regarded the tiger as a supernatural being. To please this powerful and mysterious beast, the people usually offered it animal sacrifices. Whenever a man was captured by the beast, the shaman ordered a human sacrifice. The chosen victim was tied to a tree trunk in the area where the tiger was spotted. When the person disappeared the offering was considered accepted by the animal. The tree at which this ordeal took place was then regarded as sacred by the community. Although this most cruel custom was abandoned long ago, some customs showing respect to tigers remain in this region. One can still see certain trees decorated with pieces of cloth and this is done by passing tiger hunters.

It is still considered unwise in eastern Siberia to pronounce the tiger's name in conversation in the belief that a tiger passing nearby may overhear it and become offended. There is still a custom that if someone comes upon the track of a tiger he must quickly offer some food for the sacred beast, recite certain incantations in its honour and then run away as quickly as possible.

Tigers have always played an important role in Chinese rituals and mythology. Hsuan Tan, the ancient Chinese god of wealth, is shown sometimes as a man riding a tiger. This god was believed to have the power to destroy all the evil spirits which could attack people. Miraculous powers have been attributed to various parts of the tiger's body. They have been commonly used both as a medicine and a charm. Especially popular were the tiger's whiskers which were believed to shorten the cycle of human reincarnation. Burned whiskers were applied as love charms. Chinese gamblers still believe in the miraculous power of tiger charms and use them to improve their chances in gambling.

There is a famous festival in Bengal called Bagh Yatra during which dancers perform disguised as tigers. Tigers' skins were also once considered to be useful for sitting upon during religious meditation.

The tigers' reputation for ferociously killing people and their ability to elude hunters caused tiger hunting to be the favourite pastime of Indian maharajas and aristocracy. There was a tradition in Uttar Pradesh that the governor should personally shoot a tiger during his first year in office. Prominent men used to kill many tigers and the highest score was probably achieved by the Maharaja of Surjure who was proud of killing 1707 tigers during his lifetime.

Many British kings, princes and dignitaries took part in tiger hunting which was organised by the Indian authorities. Since it is very difficult to kill a tiger in the forest, it is said that a secret trick was often played in order to please their guests. For safety, elephants were usually used for hunting, shooting from the huge animals' backs, and a professional Indian hunter was often discreetly located behind the guest. When the tiger appeared in view, the hunter would fire almost simultaneously with the guest so that one shot was heard. It was, of course, the 'second' shot which was effective as the guest-hunter usually missed, but he would be proud for the rest of his life of his fantastic shot and huge trophy from India.

Although tigers are among the most dangerous of beasts, they are not man-eaters by nature. It is usually an old or infirm animal which can no longer chase a strong prey which turns to a weaker human prey. Once a tiger acquires a taste for human flesh, however, it may become an 'addict'.

One such tiger may kill as many as a hundred people in the same region. The animal will usually chase women and children, as if it knows that they do not carry guns and are easy prey. There have been cases in India where the terror caused by an attacking tiger forced people to temporarily leave their homes and whole villages have been deserted. Tigers used to be so numerous in India that they were declared a pest and a special bounty was paid for them. For instance, 60 rupees was offered in Mysore State for one tiger killed, which was an excellent offer for it was no less than the annual salary of a forest guard. As a result many people became professional tiger hunters. But chasing tigers is very difficult because the animal is extremely alert, particularly a man-eater whose ability to fool its persecutors is remarkable. For this reason, the hunter often tried to trick the dangerous beast by disguising himself in female's clothes to conceal a gun. The hunter often had to wait many hours before the tiger decided to attack.

Tigers are still dangerous killers in India. In 1985 in the forests of West Bengal 22 people were devoured by tigers in just three months. Most vulnerable to tiger attack are fishermen, woodcutters and honey collectors in the mangrove forests of the Ganges and Bramaputra.

To deter the dangerous beasts, electrified human dummies are placed in fishing boats to lure the tigers. It is believed that when the attacking tiger touches the dummy the electric shock is so powerful that it teaches it a lesson. In fact, the number of tiger attacks has diminished in the areas where the dummies have been used.

Tigers are the largest wild cats in the world. The largest among them is the Siberian tiger (*Panthera tigris altaica*) which is regarded as a tiger subspecies. This is a real giant as it can be up to 2.8 m long and weigh up to 400 kg. These fantastic cats live in the northern regions of Asia. They have very thick coats which protect them well from the harsh winter conditions.

The tiger is a great lover. It can copulate 20 times a day. For this reason, in India, the penis of a tiger is regarded as one of the best aphrodisiacs.

Cursed House of Montezuma

Despite the general fear of the deadly rattlesnake (*Crotalus* species), for some North American Indian tribes these snakes are a symbol of life and sexual power. There is an amazing myth related to the origin of this snake among the Algonquin tribe. According to this myth, the mighty Manitoa, who was once hunting in the forest, became very irritated with his penis because it 'got in his way'. It angered the god so much that he tore off the organ and threw it away. When the penis landed on the ground it miraculously changed into a rattlesnake. Since then, this snake has been a symbol of life and sexual power in these regions.

Rattlesnakes achieved such a level of respect in ancient Mexico that humans were offered to them as sacrifices. When a Spaniard named Bernard Diaz del Castillo visited the famous court of Montezuma, he was horrified by what he witnessed there. 'They have in their cursed house many vipers and poisonous snakes which carry on their tail things that sound like bells. These are the worst vipers of all and they keep them in jars and great pottery vessels, with many feathers. There they lay their eggs and rear their young, and they give them to eat the bodies of Indians who have been sacrificed.'

There is a common belief among many North American Indians that the rattlesnake is the god responsible for bringing rain. During a prolonged drought special celebrations are held and they pray to the snake-god for rain. The most famous is the celebration in honour of the snake-god of the Hopi tribe. Thousands of people, including overseas visitors, come every year to witness the unusual festival which has no equal. The festival is well prepared, with the villagers making many trips to collect as many rattlesnakes as possible, which they take into the temple with chanting and drumming. The most spectacular event in this ceremony is the so called 'snake dance'. This frightening celebration is performed by the priests of the tribe. One after another, the priest-dancers perform their dances, holding a rattlesnake firmly between their lips. After this horrific dance, watched by an excited crowd, the snakes are gently removed and put on the ground, from where they are carried back to their forest homes with great respect.

As a result of such an unusual 'dance' the long expected rain is said to follow soon. Surprisingly, it is said that during the ritual dance the deadly snakes usually 'behave' as none of the priests or their assistants have ever been bitten. The snake worshippers explain that these snakes are friendly towards men because they symbolise a god. But a herpetologist who witnessed the strange snake dance claims that the rattlesnakes behave so well because they have their deadly fangs secretly removed before the event.

Rattlesnakes are also believed to possess the power to relieve pain. Dried and powdered poison fangs of the snake are used for pain relief during childbirth. The skin of the rattlesnake is used for similar purposes. Women cover themselves with the snake's skin in the belief that it will ease the pain during delivery.

One of the main reasons that these snakes are so highly venerated is that they they possess a strange 'rattle' at the end of their tails, which does not occur in any other snakes. The rattle is made by interlocking segments of keratin which resemble buttons. When the snake vigorously shakes its tail, the 'rattle' vibrates, making a distinct sound. This sound is likened to the hiss or rapid cracking of frying oil. The number of the keratin cups indicates the age of the snake since a button is added every time the snake sheds its skin and this may take place four times a year. Some people say that the rattlesnake is a real gentleman as he 'rings' his 'bell' to warn his attacker to retreat before using his deadly weapon — his poisonous fangs. Biologists have tried to explain the reason for the existence of this strange rattle by claiming that it is an organ of defence. One biologist, however, who studied these snakes was surprised when he noticed that the rattle was only used by 4 per cent of the observed snakes. It still remains a mystery why nature has invented such a sophisticated device if it is only used occasionally or maybe not at all.

Biologists were also fascinated by the remarkable ability of rattlesnakes to detect their prey in complete darkness. An interesting experiment was performed in the USA. A rattlesnake, which was kept in a cage, was treated with odorous chemicals to eliminate its sense of smell and had its eyes covered. A live mouse was put into the cage. Despite this sensory deprivation the rattlesnake located and caught the mouse with ease. It was found that rattlesnakes have a special adaptation. Two pits between the nostrils and eyes enable the snake to detect the slightest temperature change in its surroundings. In this way the snake can quickly detect prey inside its burrow in total darkness without using its eyes. It was later found that certain other snake species also have this adaptation.

Blood-Thirsty Frog Goddess

In the past, frogs were held in great esteem in various regions of the world. In ancient Mexico a frog-goddess called Centeotl was regarded as the patron of childbirth. This goddess was among the most powerful and demanding of deities. To please her it was customary to perform human sacrifices in temples erected specially for her worship. To emphasise the femininity of the deity, it is said that only females were offered to her at the sacrificial altars.

The ancient Egyptians had a goddess called Heket who was depicted as a woman with a frog's head or even simply as a frog. At a certain stage in the history of ancient Egypt this frog-goddess and her husband were thought to be the first to 'build men and make gods'. It was believed that the goddess Heket was always present at the birth of the pharaoh's babies to breathe life into the newly born. No wonder that Egyptian pharaohs made generous offerings to please this important frog-goddess. Interestingly, the Egyptian hieroglyph for human embryo was the shape of a frog. The frog was also used to describe a very large number such as 100,000 and this choice of symbol was associated with the enormous number of frogs which appear after rain. In many parts of Africa frogs were linked with fertility beliefs and for this reason frog amulets were worn and some Africans even ate frogs believing it to be a powerful aphrodisiac.

Some people in India believed that the world rested on the body of a giant frog, and that the frog's movements were the cause of disastrous earthquakes. The Chinese used to think that the lunar eclipses were also caused by a gigantic frog which swallowed part of the moon.

Even now the frog is the object of veneration in some regions of Malaysia. Here a special frog *Rhacophorus reinwardtii* has been chosen as sacred, because of its remarkable ability to glide from trees. In honour of the sacred frog special ceremonies are organised which appear very odd to strangers. During a peculiar street parade a procession of people carry their 'sacred' frogs placed comfortably in chairs specially constructed for this occasion.

Aborigines living in Australia at Lake Tyers also believe in the mysterious power of frogs to bring rain. According to their myth, there was once a drought which was so severe and prolonged that both people and animals living in the desert were about to die from thirst. The animals then decided to meet and discuss what to do. One of the animals, the story says, declared that the cause of the drought was a gigantic frog which had swallowed all the water. After discussion, the animals agreed that the only remedy would be to make the giant frog laugh as this would cause her to disgorge the water from her stomach. All the animals tried to make the frog laugh but in vain. Fortunately there was an eel among them. When this fish started to wriggle and distort itself the giant frog suddenly began to laugh and when she opened her mouth the water flooded over the 'thirsty land'.

The Australian Aborigines who live in the desert regions where there is sometimes no rain for a number of years occasionally 'drink' water from frogs' bodies. There are some frogs which absorb water during the short period of rain to such an extent that they become almost oval. The frogs then make a burrow in the sand in which they live for up to ten years, until the next rain comes. In order to stop water evaporating from their skins, they produce a kind of membrane from their skins which keeps them water-tight. The Aborigines simply dig out one of these barrel-like frogs and, by squeezing them, have sufficient liquid to quench their thirst in the heat of the desert. The so-called Australian water-holding frogs belong to the genus *Cyclorana.* These central Australian frogs appear in such enormous numbers in certain areas where the breeding conditions are right that they have created problems. Occasionally, in the process of migrating, the frogs cross railway lines and so many frogs have been smashed by trains that the tracks have become unusable for some time because they are so slippery.

2.
Animals in Magic and Rituals

Blood Brotherhood

The leopard (*Panthera pardus*) is the most dangerous of all wild cats in terms of maliciousness and savagery. Although it is smaller than the lion or the tiger, people are usually more afraid of the sudden attack of the leopard.

The fear of the ferocious leopard is so strong in Africa that this animal has become the totem of certain tribes. In some parts of West Africa a blood brotherhood has been established between this deadly beast and men. The inhabitants of Naga Hill treat leopards with religious respect. They do not kill them and in order to establish a closer relationship with this animal, they must capture one alive. Once the animal is gently bound, an incision is made to obtain some of the animal's blood. During a special ceremony the leopard worshippers mix the animal's blood with their own and drink it, to establish a kind of blood-brotherhood. The animal is then fed well and released. The leopard's brothers are proud of their relationship with the powerful beast and they imitate the animal's appearance by cutting whiskers on their faces. This blood relationship is so strong that if the leopard dies his human bloodbrother should also die. Fortunately, the human brother never finds out about his brother leopard's death, so he continues his life, always believing his brother leopard is alive.

This is not, however, the only strange association of man with the leopard. In Western and Central Africa, especially in Sierra Leone, there used to be 'Leopard Societies'. These societies are still said to exist in some remote jungle regions, but are kept secret. It was not easy to become a member of such a society and those who were accepted were given a pronged knife called the 'leopard knife'. This was the weapon to be used to commit murders if ordered by the society.

The leopard man imitated the beast by covering himself with the animal's skin and painting his face with spots. After attaching metal claws to his hands the leopard man was ready, when ordered, to attack. He started to behave like a ferocious beast and did not even seem to recognise his closest family members. If ordered, he was ready to kill them indiscriminately. After chasing them, he killed a man, woman or child using the powerful, pronged knife and then he ritually dismembered the body. He cut out the heart, liver and intestines of the victim and cut the rest of the body into pieces, so that it would be unidentifiable when found in the forest. To fool the people, the leopard man even left leopard footprints on the soft ground by using a specially made wooden club.

The details of the ritual slaughter differ depending upon the society and sometimes the victim is tortured before his death. In the judgement of the leopard men these ferocious killings of innocent people are morally justified. Ritual murders are committed when the unity of the tribe is threatened, for instance, when there is rivalry over succession after the death of a chief. Once the villagers, frightened by the repeated deaths, decide to stop quarrelling and make the choice of the new chief, the 'secret' murders immediately cease.

It is said that the leopard men only commit their murders in response to their ancestors' command, whose desire is to stop the weakening and possible disintegration of the tribe. The leopard men quickly become normal tribe members and return to their family. It is said that they may even forget what they have done.

In many regions of Africa leopards are believed to have miraculous powers. Zulu witchdoctors in South Africa are said to use a leopard's skin instead of a crystal ball to forcast the future.

Not only are the relations between men and leopards fascinating, but the animal itself is also very interesting. This beautiful wild cat is an astute hunter. It ambushes its prey by lying hidden on a tree branch from which he drops directly onto a passing animal. In order to catch its prey the leopard uses a very shrewd ruse.

When he spots a deer he starts rolling on the ground as if taking no notice of it. The curious victim comes nearer and nearer to observe its playful activity. The leopard then strikes as swift as lightning and catches its easy prey.

The leopard likes to consume its prey in a quiet place free of scavengers so it sometimes carries the carcass to the nearest tree. It is so strong that it can carry a big antelope or even a young giraffe and place it on a horizontal branch. After a meal, which may take two or three days, only the animal's bones and pieces of its skin may be seen hanging from the tree.

The leopard seems to realise that buffaloes have a good sense of smell and can detect its presence from a distance. To prevent this, the clever leopard rolls in fresh buffalo dung and so can approach and kill his prey.

Not all leopards have the characteristic black spots on their bodies. Certain leopards found in other regions, such as Malaysia, have entirely black fur. These interesting animals, which were formerly considered to be a distinct species, are in fact a variety of the same species. They are commonly known as black panthers.

Mysterious Soul

For inhabitants of the northern regions of the world, especially Eskimos, seals are the most important animals. These bulky animals used to provide them with both food and dress. The meat was their staple food and the oil from the fat deposits was used for lighting their homes. Even canoes were made from seal skin stretched over their bones. Although Eskimos had to kill many seals in order to survive, they showed great respect towards these valuable animals.

Incredible rituals are associated with seals. Among the Iglulik people, catching his first seal is an important event in the life of a young man. To celebrate, the boy lies on the ice, and his father hauls the dead seal across the boy's back in the belief that as a result his son will have good luck in his future hunting expeditions.

There is an even more amazing custom related to seals among some Eskimos. When the seal is killed the hunter's wife pours some water on the animal's snout. This is done in the belief that the water will relieve the animal's thirst and hence prevent the revenge of the animal's soul, which is claimed to be alive.

The belief in the importance of appeasing the killed seal's soul is so strong that certain other surprising rules are observed by some Eskimos. The harpoon which was used to kill a seal during an expedition must be stood beside a lamp in which seal blubber oil is burned. It is strongly believed that the soul of the killed animal inhabits the harpoon and that it must have plenty of comfort by resting for some time near the flame of the lamp. A truly amazing belief.

Inhabitants of West Greenland have another custom. They store the heads of all the killed seals in a huge pile in front of their huts in order to show their respect towards the animals' souls. There is a custom in Labrador that when a seal is killed the tip of the animal's heart and the liver must be thrown into the sea. This is done in the belief that it will ensure luck in future seal hunting.

Interesting festivals, known as 'the sinking of the seal bladders', are held among the Eskimos in the Bering Strait. During this event the bladders of all the animals killed during the year are ceremonially sunk in the sea through a specially made hole in the ice. This ceremony commemorates a particular seal species, the bearded seal. Shamans play an important role during this celebration, praying for a good harvest in the coming year.

The most interesting of the many species of seal is the northern fur seal (*Callorhinus ursinus*). This animal occurs in very large numbers in certain regions. It has been estimated that up to two million of these large animals can congregate around a small island in the Bering Sea, and this is probably one of the largest concentrations of migrating sea mammals in the world.

Male northern fur seals establish harems on their territory. One adult bull seal may have a harem of up to 50 females. Only the strongest males can form a harem and the strong competition means that only about 10 per cent of bulls can find a sexual partner and mate. Once the harem is established the 'king' male constantly guards his harem for the whole breeding season and does not have time to look for food. When the breeding season is over the 'owner' of the harem is often so weak that he hardly has the energy to return to the sea to feed.

To escape the harsh winter conditions these seals migrate southwards. It is an extraordinary migration as the animals can cover some 10,000 km swimming to Japan from their breeding grounds near the Bering Sea, then to California and then back. This enormous trip, during which they never touch land, takes about eight months. Only the females and young make this long trip. The 'lazy' males only travel some 600 km.

The largest of all seals is the southern elephant seal (*Mirounga leonina*). The strange name is most appropriate since it is a very large animal and can be as long as an elephant. In addition, the male adult seal has an extraordinary elephant-like trunk which inflates to some 60 cm during the rut.

African Hero

Shrews are among the most common mammals in the world. There are about 250 living species and they occur in almost all parts of the world including polar regions. These small, nocturnal rodents were regarded in ancient Egypt as symbols of night and darkness and were held in great respect. They were buried in special tombs and were often mummified – many mummified shrews were discovered at Thebes.

In contrast, Eskimos fear these tiny animals and dread crossing their path. They believe that a shrew will always attack a man and after burrowing through his flesh will stab his heart with its elongated head, inflicting a horrible death.

In some parts of Central and East Africa, a certain species of shrew became so famous that it was called the African Hero. It is the armoured shrew (*Scutisorex somereni*), a small animal, about 15 cm in length. This animal is greatly admired for its remarkable strength and it became a peculiar artist in a fascinating show of strength, especially in Zaire and Uganda. In this unusual show the shrew is placed in the middle of a stage, surrounded by the audience. First, everyone prays and chants and then a strongly built man steps barefooted upon the back of the animal and balances in this position on one foot for about five minutes. The man may weigh some 90 kg and any other animal the size of this shrew would certainly be crushed to death. But not the African Hero. The animal moves off quite unperturbed, to the amazement of the watching crowd.

Although this shrew seems to be no different from other species, it has a unique strength. It has more than twice the number of vertebrae that other shrews have. These vertebrae are very firmly joined and so make an incredibly powerful spinal column, which gives the animal a great advantage as it can crawl under heavy rocks in its search for food.

The Mangbetu people of Zaire believe that this shrew possesses a miraculous power which can be transferred to people. The charred body of a shrew is used as a charm to bring strength. Wearing such a talisman is believed to give the wearer special power during a fight against his enemies. Some locals in central Africa used to carry this charm when setting out on an elephant hunt, believing that it would make them invisible to the elephant.

These small animals, the size of a common rat, are famous for their unusual metabolic rate. The shrew's pulse rate is incredible and when frightened it may be over 1,000 beats per minute. The animal must consume an enormous amount of food to survive and it can eat three times its body weight of food in one day. As a result, its life expectancy is extremely low. These shrews usually die after a year and a 15-month-old shrew is regarded as an exceptionally old fellow.

Despite their small size, these animals are known to fight viciously. In some regions, the fight between two shrews put in the same cage is a popular show. When a piece of meat is given to the two shrews in the cage they start fighting, screaming and biting each other with incredible ferocity. At the end of this rather mad show the victorious animal will devour its foe.

There are some shrews which have another weapon in addition to their fighting prowess. The short-tailed shrew (*Blarina brevicauda*) has glands in its mouth which secrete poisonous saliva and a single bite is sufficient to kill a small prey in just a few seconds. This shrew is unique since no other mammal has a poison gland in its mouth. This shrew lives in North America and is about 10 cm long when measured without its tail.

The smallest non-flying mammal in the world is a shrew. It is the Etruscan or Savi's pigmy shrew (*Suncus etruscus*) which is only about 3.5 cm long and weighs about 2.5 g when adult. It is such a tiny mammal that it can easily crawl into the narrow hole made by an ordinary earthworm. It is also an excellent jumper since it can jump up to 20 cm and this is more than five times its own body length.

Shaman's Friend

The polar bear (*Thalarctos maritimus*) is a most impressive animal and was worshipped some thousands of years ago by Neanderthal man who hunted them and was often killed by these giant beasts. With its long, steel-hard claws, the mighty animal must have been a most formidable enemy for primitive hunters who only had flint-tipped spears to fight with.

The polar bear is the world's largest quadruped, carnivorous animal. The male can be up to 3 m tall and weigh up to 650 kg. The largest polar bear ever shot was 3.3 m in height and weighed 1002 kg. These exotic monsters were famous in ancient Egypt. A polar bear was brought to Egypt from the north on the orders of king Ptolemy II (285—246 BC) and was kept in his royal zoo. This animal was regarded as a symbol of strength and power and took part in important ceremonies. The giant polar bear led processions through the streets of ancient Alexandria. A group of men, it is said, followed the polar bear carrying an enormous gilded phallus, some 50 m tall.

There are records that polar bears were also used for demonstrating the skill and bravery of warriors in ancient Rome. In 237 AD, during a special celebration in the Roman Coliseum, some 50,000 spectators watched gladiators in a bloody contest with 1000 polar bears imported for this incredible show. This strange event was sponsored by a wealthy citizen of Rome who later became Emperor Gordian I.

To own a live polar bear used to be a great privilege. When an Icelander named Isleifr wanted to become a bishop, he sent a polar bear as a gift to the Holy Roman Emperor Henry III in 1054 AD. The emperor was so delighted, the story says, that the man obtained his nomination. Similarly, in 880 AD Ingimundr the Old caught two polar bear cubs and offered them to King Harold the Fair-haired of Norway, who was so delighted that he rewarded him with a ship full of precious timber.

The impressive polar bear was held in highest esteem, however, by the inhabitants of the Arctic regions. The Eskimos used to believe that this animal was the supreme being and could understand human language. They even believed that this animal could change itself into a man, and after his death the man would change back into a polar bear.

Despite their admiration, Eskimos hunted polar bears extensively since they provided them with meat and fur. As it was very dangerous and difficult to kill this powerful beast, those who were brave enough to do so acquired great prestige in the community. Polar bears can run exceedingly fast and can skilfully drift away on floating ice and many hunters were killed by these ferocious beasts. Accordingly, when a polar bear hunt was successful, special ceremonies were held to conciliate the bear's soul. It was believed that if the soul of the dead animal was offended it would depart in anger, and evil might strike the village. To prevent this, the bear's skull had to be taken to the communal house and placed in honour upon a raised bench. The skull remained there for some time and was offered various gifts. Then the 'polar' bear dance took place. This dance was an extravagant performance and the feast took several days. After the dance was over the bear's skull was placed on an ice floe and when it moved it was believed that the animal's soul had shown its willingness to depart.

Some Eskimos believed that by wearing polar bear teeth they could acquire the animal's power and strength. Moreover, it was believed that a man who desired to become a shaman could only increase his power by asking the polar bear for help. He had to travel alone on a floe until he met a large herd of polar bears. He would be almost frightened to death, but if he fell upon his face to show respect, one of the mighty bears would step forward and ask him if he wanted him to become his magic helper. The man would, of course, say yes and the polar bear and the man would then return together to the Eskimo's house. They became such good friends that the bear, the story says, caught a seal on the way and offered it to the man. The man then became a shaman and would be as strong as his assistant from whom he could always expect help.

Polar bears are the only large land carnivores which are never afraid of man. A polar bear will kill and eat a human hunter and never gives up a fight, continuing to charge even if seriously wounded. A polar bear was once recorded charging a man for over 50 m even though he had several bullets in his heart. Polar bears are also vengeful animals. They are known to destroy whole igloos with women and children inside when the men are away on a polar bear hunting expedition.

The body of the polar bear is so densely covered with white fur that only its black nose breaks this perfect camouflage in the white surrounding of the Arctic. The intelligent animal seems to know this as it often covers its nose with snow or its paw to make it invisible during hunting expeditions.

He is a cunning hunter and uses a special strategy to catch seals. He waits at a hole in the ice where the seal will eventually emerge to breathe, and he closes the other holes with ice and snow to make sure they cannot be used by the seal. When the seal emerges, the bear kills it with a single blow of his paw.

Throughout the ages, the mighty polar bear has been attributed with unusual healing power. An ointment made from its fat is used as a cure for rheumatism. In Korea the most medicinally valuable is the animal's gall bladder. It is in great demand as the best cure for liver and bile-tract disorders. The enormous sum of nearly $3,000 was once paid for a single polar bear bladder.

An interesting story is related to the liver of the polar bear. Scientists long ago were intrigued by the fact that Eskimos, who eat the flesh and fat of this animal as their main food, always discard the liver. Recent investigations have established that the Eskimos were right in their attitude. The liver of a polar bear contains an incredibly high amount of vitamin A, up to 30,000 units in one gram, so that eating such a liver would be hazardous. In overdose, this important vitamin becomes a deadly poison.

3.
Sacred Creatures

Sacred Rats

Since ancient times, the rat has been both hated and loved by human beings. The ancient Egyptians regarded the rat as a god which, surprisingly, symbolised both total destruction and wisdom. During the Middle Ages, Europeans believed that rats were created by the devil. It was also believed that witches changed into rats in order to move about unnoticed and to get secretly involved in the affairs of humans. For the Japanese, the rat was a messenger of the god of wealth and it was believed that if rats ate the New Year's rice cakes, there would be a good harvest that year.

The highest level of veneration for rats has developed in India. The Hindu god Ganesa, a symbol of prosperity, is traditionally believed to be carried by rats. In India one can often find shrines holding the golden, rat-borne Ganesa. One of the Hindu sects in Rajasthan has a special temple dedicated to the worship of rats. In this particular temple one can see thousands of rat inhabitants fed daily with a mixture called 'laddu', which consists of grain, sweets and milk. The sacred rats have no fear of humans and a large number of them move around the worshippers in the temple.

Curiously, even in Europe the rat was associated with pagan religion. For instance, before Christianity, the Germans used to worship a goddess called Hludane, whose symbol was the rat.

The Chinese considered the rat to be a great delicacy and during the T'ang dynasty (618–907) newborn rats stuffed with honey were served as food. According to one report, the newborn rats slowly moved about the table crying feebly and they were picked up with chopsticks and eaten alive. The early Chinese hunted rats and called them 'household deer'. Rat hunting was also common on some Pacific islands. In his book published in 1880, W. W. Gill said: 'In those days — ere the cat had been introduced — rats were very plentiful. Rat hunting was the grave employment of bearded men, the flesh being regarded as delicious.' On the Tonga islands rat hunting became a popular sport and in about 1806 the king and his entourage shot the rats in the forests with 2-metre-long arrows. As J. Martin noted in 1818, 'whichever party kills ten rats first, wins the game. If there be plenty of rats, they generally play three to four games.' Rats were also an important source of food on Easter Island, as Captain Cook wrote in 1774: 'they have also rats, which it seems they eat; for I saw a man with some dead ones in his hand, and he seemed unwilling to part with them, giving me to understand they were for food.'

During the nineteenth century in London, rats were involved in a popular form of gambling. There were special matches in which a number of rats were held in a small arena into which dogs were released to catch and kill as many rats as possible in a given time. The spectators bet on which dog would win. The best dog was said to have killed 500 rats in just five minutes. This 'sport' was so popular that one house in London had to purchase 500 rats a week for the show. Rat catching became a good source of income for professional catchers who supplied the sporting houses.

Rats have an enormous capacity to survive. When a rat is dropped from a five storey building it survives unharmed and quickly runs away to safety. It will survive when flushed down the toilet, and by using this route it can unexpectedly re-enter the house. With its remarkable teeth a rat can gnaw lead pipes and cinders. Its body is so flexible that it can pass through a hole no larger than 2.5 cm and rats can climb high, smooth walls as if they were ladders. Rats have an incredible resistance to radioactive contamination. In 1980, a rat population was discovered on the island of Runit in the West Pacific. These rats are descendants of an underground rat population which survived 43 atomic bomb tests carried out there in the 1940s and 1950s. Because of these atomic tests, the island of Runit is regarded as unsafe for humans for at least the next 25,000 years.

The rat is a very intelligent and sometimes even compassionate animal. A miner was once feeding a friendly rat in a mine, when it suddenly ran away with a piece of rope in its mouth. He soon returned, still with one end of the rope in its mouth, while at the other end there was a blind rat.

Rats are even known to save human lives. According to one nineteenth century account, a clergyman kept a rat as a pet in his home. One night when the clergyman was asleep, the curtains in his room caught fire. The friendly rat responded immediately, biting his friend on the cheek and alarming him in time to escape unhurt.

Few people realise that there are probably as many rats in the world as there are people and these rodents cause enormous damage to crops. In Asia alone the rat population consumes enough rice to feed more than 250 million people annually.

Rats, and the many parasites which live on their bodies, carry at least 20 different deadly diseases, including typhus and trichinosis. In medieval Europe these rodents spread the deadly plague called the Black Death. The rat flea is parasitic on black rats (*Rattus rattus*) and when the infected fleas bite people they transfer the dangerous plague bacteria.

The Strangest Clock

The hamadryad baboon (*Papio hamadryas*) must have profoundly impressed the ancient Egyptians since it was held in great respect and was even regarded as sacred. A baboon god called Thoth was a lunar divinity and a patron of artists and physicians. This god was also responsible for governing the passing of time. Consequently, an image of the hamadryad baboon was used in a most unusual way. An Egyptian clock was made of a sculpture of a seated male baboon, but, instead of a clock pointer, water flowed from the baboon's phallus to mark the passing time. The Egyptians believed that baboons have a remarkable habit of passing water precisely every hour of the day and night, hence the idea of this particular clock. The male hamadryad baboon was also considered to have a great sexual power and was frequently depicted in Egyptian art as a symbol of masculinity, often in a state of sexual arousal.

These baboons were admired for their high intelligence and the Egyptians even believed that their famous hieroglyphs were not the invention of man but of an intelligent baboon. The respect shown to this monkey was so surprisingly high that when a hamadryad baboon died it was embalmed like a human aristocrat and funeral ceremonies took up to two days. Archaeologists discovered a large baboon cemetery in the famous Valley of the Kings in Egypt.

Baboons, of which there are five species, were also worshipped in other parts of Africa. The members of the Baule tribe of Ivory Coast had a baboon-god called Kbekke, who was portrayed as a man with a baboon's head. The statue of this god had a special bowl in front of it and offerings were brought and placed in the bowl. To please the god, even sacrificial blood was occasionally included in the offerings.

Baboons were well known in ancient Rome and there are reports that these animals were imported from Africa in the second century BC and were kept as domestic pets by aristocrats. They were occasionally treated like humans and dressed in extravagant and costly clothes. It is claimed that they were even taught to 'play' musical instruments, such as the flute.

There are many stories about the remarkable ability of baboons to perform certain tasks such as collecting fruit in the garden, but the job held by a baboon in South Africa was probably the most remarkable. A station master who owned a tamed baboon one day lost his leg in an accident. As a result, he was unable to move quickly enough to operate the levers located outside the railway station, which guided the signals. The skilfulness of his pet became a wonderful solution to the problem. The station master was able to teach the baboon what to do, and the clever animal learned to follow his instructions. He knew when and how to pull each lever in order to direct a train to the right rails. The master only had to call the number of the lever and the baboon responded immediately. He never made a mistake.

One intelligent and famous baboon became a soldier in the army.

Dressed in uniform, the baboon was an excellent sentinel. He was able to immediately detect the approaching enemy and by a distinct, short bark informed his commander of the danger. His hearing ability was incredible. He is probably the only animal in history to reach the rank of corporal. Whenever an officer passed he immediately stood to attention and saluted him with respect.

Even in the wild the baboon shows a high degree of intelligence and may occasionally use simple tools. When a baboon finds a scorpion he breaks off the poisonous sting from its tail with a stick and then eats the remaining flesh.

Although a baboon may become a close friend of man, in the wild he may be dangerous and, if molested, may severely wound a man. There are stories about baboons raiding huts and stealing sleeping infants. It was believed that these babies were devoured, but some scientists suggest that baboons may steal human babies and 'adopt' them as their own.

The baboon has powerful teeth and is a fearless fighter. When in groups, baboons can deter such ferocious beasts as leopards. Males are known to fight their rivals over the possession of a female and the struggle can be so brutal that the female, snatched from one male to another, may be seriously wounded. Even if a female dies, the victorious baboon considers her to be his exclusive property and will fight if someone tries to take her body away from him.

The male hamadryad baboons usually have a harem of up to ten wives. The overlord, who is bigger than the females, keeps them in line. He seems to be very 'jealous' and he can even kill a female if he finds her breaking her bond by mating with another male. He regards himself, however, as above these rules since, if he is strong enough, he will steal a female belonging to another male and mate with her. A hamadryad baboon troop may consist of up to eighty individuals. The troop is always under the strict command of the strongest males. The baboon group, like a military unit, maintains a constant guard. The chosen animals are ordered to stay on the hilltops and warn the other members if a dangerous predator is spotted. When a group moves, the weaker females and their young are always kept in the centre with the stronger males on the outside to defend the group in the case of an attack. It is known that the strongest males take their responsibility seriously as they often risk their lives defending their community.

Sacred Giant

The Asian elephant (*Elephas maximus*) has played an important role in the rituals and customs of native people since ancient times. Ganesa, the famous Hindu deity of fortune and wisdom, has the body of a man but the head of an elephant. For Hindus, Ganesa presides over the beginnings of all projects such as the opening of a new business, the starting of a journey or the building of a new house.

In Thailand the highest level of veneration was attributed not to every elephant, but to a specific and extremely rare white elephant. An albino elephant was considered a sacred animal and the appearance of such an elephant was the most important event in the whole region.

Whenever the governor of one of the provinces learned about the appearance of an albino elephant in his area, he immediately organised an expedition to find and bring the precious, sacred animal to the king's court. He ordered that prayers be offered in all temples in the sacred elephant's honour. A wide path was specially cleared in the forest so that the great guest could easily reach the capital.

The elephant was escorted the whole way by numerous servants who played musical instruments and sang for his amusement. According to one report, 'the king and his court, all the chief personages of the kingdom and a multitude of priests accompanied by troops of players and musicians came to meet him...' The white elephant was given a special pavilion to live in and the animal was dressed in a superb purple cloak richly adorned with golden ornaments. The sacred animal was served by special slaves and when he was having a bath a high ranking officer had to hold up a huge golden umbrella to shield his head. The court always reverently bowed before the sacred animal offering him homage.

When the sacred elephant arrived at the new pavilion in the court, he was greeted and then asked in a special speech to accept the new accommodation: 'O father Elephant! We entreat you to banish every wish to stay in the forest. Look at this delightful place, this heavenly city! It abounds with wealth and everything your eyes could wish to see or your heart desire to possess. It is of your own merit that you have to behold this beautiful city, to enjoy its wealth and to be the favourite guest of His most exalted Majesty the King.'

Sometimes the death of a white elephant causes incredible turmoil in the country. In 1862 a white elephant was found and the news about it reached King Rama IV. The delighted monarch immediately ordered the construction of an extravagant building to receive the guest. But before the 'palace' was finished the 'sacred' elephant died. According to Anna Leonowens, who described this event, nobody was brave enough to reveal the terrible truth to the king. But the one responsible for building the pavilion decided to stop the work. When the king arrived and saw no one working he realised the truth. 'With a cry of pain he sank down upon a stone, and gave vent to an hysterical passion of tears.'

Although such a degree of veneration seems incredible, even as late as the 1920s a white elephant was treated with enormous respect in Thailand. When an albino elephant was found in 1922 the animal was transported to the king's palace in a specially adapted train with a railcar equipped with an electric fan and a shower to make the elephant's journey as comfortable as possible. The sacred animal was then led with great splendour through the streets of Bangkok. In the palace the king greeted his admirable guest and personally anointed his body with precious perfumes.

If a young white elephant was found it was also treated with great care — if it was very young it would be suckled by numerous women.

Insect of Immortality

No other insect has ever achieved such a high degree of veneration as the humble scarab beetle (*Scarabaeus sacer*), which feeds on dung. For the ancient Egyptians, the balls of dung made by these small creatures became the symbols of the sun god, the chief Egyptian god during the New Kingdom. Burying and then retrieving the dung ball from the ground ritually represented the rising and setting of the sun.

It was erroneously believed that there were only male beetles and that they could miraculously reproduce without union with the opposite sex. Because the young beetles were observed to emerge from the dung ball buried in the ground, the scarab beetle also became the symbol of immortality and an essential part of the various mortuary rituals. To ensure immortality, it was customary to remove the heart of the dead and replace it with a scarab beetle or its effigy. Carved in precious stone, the scarab beetle was also placed on the breast of a mummy. It was believed that the presence of the sacred beetle would not only secure the best judgement of the soul but would bring forgiveness of the man's sins and assure immortality. A scarab beetle made of precious stone and set in gold was discovered in the famous tomb of the pharaoh Tutankhamen.

In ancient Rome the famous Egyptian beetle was used as a charm to bring luck. It was customary for Roman soldiers preparing for battle to take a scarab beetle amulet with them in the belief that it would protect them from the enemy soldiers' weapons.

The belief in the magic powers of the scarab beetle has survived till recent times. Some Egyptian Beduins still use scarab beetles in their initiation ceremonies. They believe that only a sacred scarab beetle can change a boy into a man. In a peculiar ceremony, the young boys to be initiated stand in a special place surrounded by adults while the sheikh in charge of the ceremony calls the 99 names of Allah and all present loudly repeat the names. When incantations reach a proper level, the sheikh announces that the ceremonial consuming of scarab beetles has to commence. Each boy gets one scarab and, without hesitating, he must quickly eat it. Most obey this unusual command and are then regarded as adults.

Dung beetles are also famous in other regions of the world. The Kaua Indians in north-western Brazil were known to perform a ritual dance called 'Dance of the Dung Beetles'. Koch-Grunberg, who witnessed such an event in 1921, described it in the following way: 'Hand in hand and singing, two masked dancers walked forwards and backward, one beside the other. They held ceremonial staffs tightly clasped under their outside arms, using them to roll back and forth a stick which represented the ball of dung.' By imitating the famous insect's peculiar behaviour, the dancers tried to identify themselves with the sacred beetle.

Because dung beetles eat dung they usually live in pastures where large animals graze. When a scarab beetle finds a pile of fairly fresh dung it forms it into a ball, which is often as large as a golf ball although the beetle itself is usually not more than some 2.5 cm long. After the ball is formed, the beetle, using its hind legs, pushes it backwards until it reaches a depression. It then buries itself with the ball in the depression and feeds on the dung until the whole ball is consumed.

At a certain time of the year, dung balls are formed for another purpose. This time the female lays an egg inside it. The ball is then buried and the hatched larva feeds on its shelter leaving only a thin outer shell. The larva then pupates inside the shell and an adult insect emerges later. It was this peculiar emergence which fascinated the ancient Egyptians so much.

Hindu Sacred Monkey

People who are not familiar with Hindu beliefs are usually most surprised when they observe certain monkeys roaming at will in the bazaars of India. Monkeys enter stores and eat or pick up food or other objects and no one harms them. On the contrary they seem to be welcomed everywhere. This high degree of veneration is given to the hanuman langur (*Semnopithecus entellus*) by Hindus living in India and Sri Lanka. Some of these monkeys are known to rob crops, while others can be seen comfortably living in Hindu temples, where they are provided with food by the local inhabitants.

The reason for such unusual esteem is that the Hindus believe in the monkey-god Hanuman who miraculously helped rescue Sita, Rama's wife, from the vicious attack of the demon Ramana. Rama is one of three important Hindu gods and Hanuman had the privilege of becoming Rama's assistant. The monkey-god Hanuman is believed to be a patron of all arts and healing. Millions of people in India worship Hanuman and his effigies hang in many Hindu homes. There are special temples devoted to the monkey-god. Some pilgrims on their way to a Hanuman temple try to show their great respect to the monkey god by imitating the monkeys' behaviour. During religious festivals, thousands of monkey god worshippers march in a very colourful and spectacular procession along the streets carrying images of the monkey god. People are very excited and as one witness said 'no one refuses to offer gifts to the lifted bowls of many beggars sitting along the street.'

Some Indians believe that a person would soon die if he decided to live in a place where a hanuman monkey has died. There are special exorcists who are asked to check that the site of the building is not 'inhabited' by the hidden remains of a hanuman monkey.

As hanuman monkeys are regarded as sacred by many people in

India, and it is a great offence to inflict harm on them, some 'wicked' people 'invite' these monkeys to harm their enemies. When someone wants to take revenge on a hated neighbour or enemy, he will throw a handful of rice on the roof of his house. The sacred monkeys soon notice what has happened and they climb on the roof to get at the rice. Since the rice usually falls into small cracks between the roof tiles, the monkeys tear at them to get the food and, as a result a substantial part of the enemy's roof can be damaged. Nobody blames the sacred monkeys.

Langur monkeys are regarded as the fastest primates on the ground as they can run up to 40 kmh. They can also make remarkably long leaps. A single leap from a treetop to a branch of another tree may be up to 15 m. It is that when a hanuman langur monkey is born, many females surround the baby, evidently excited by the new arrival. The infant is an entirely different colour from the adults. The highly respected hanuman monkeys are known to commit infanticide. This happens when in a mixed troop the breeding male is replaced by a stranger male who may kill all the females' infants. This unusual behaviour is explained by scientists who claim that the new breeder is aware that a female who loses her offspring becomes sexually receptive sooner as she does not need to produce milk to feed her baby.

An interesting monkey which belongs to langurs is the proboscis monkey (*Nasalis larvatus*) found in swamp forests of Borneo. Some biologists regard it as the oddest-looking monkey in the world. The male has an extraordinary nose, which can be up to 17 cm long and hangs down its chin. Although the reason for having such an unusual nose is not clear it seems that it helps the male to make its deeply resonant 'honking' call. The locals call this monkey a 'blanda', meaning Dutchman, a name which refers to early Dutch colonists.

Monkeys are also a familiar sight in some temples in Nepal. Here, however, it is the rhesus macaque (*Macaca mulatta*) which is the object of worship. Large groups of these monkeys can be seen freely walking around some Hindu temples. It is said that these sacred monkeys were invited to the temple by people some 2000 years ago, and have been living there ever since, generation after generation. Today some 300 rhesus monkeys live permanently in the famous Hindu temple at Pashupati in Kathmandu, the capital of Nepal. The sacred monkeys feel at home in this important Hindu shrine and they are fed daily with rice, peanuts and pumpkins. They may leave the temple and walk around freely and people offer them candy and treat them with great respect. These sacred monkeys are strictly protected by law in Nepal.

Rhesus monkeys have also found their home in a Buddhist temple at Swayambhu since they permanently live around the temple. It is assumed that the monkeys came from the forest long ago and when people started to give them plenty of food, they decided to stay and left the forest forever.

Symbol of Fertility

When the early Portuguese sailors returned from their voyages to India, they brought back stories of a most dangerous serpent which they called 'cobra de capello', meaning 'serpent of the hood'. They were describing the cobra.

There are a dozen species of cobra and all are recognised by their hoods. When alarmed, a cobra first demonstrates its deadliness by hissing loudly and then by raising a third of its body upright and flattening the skin of its neck to form a wide hood. As a result, the snake's head looks much larger than it really is and makes it much more frightening.

The hood of the Asian cobra (*Naja naja*), is ornamented with 'spectacle-like' markings which makes this snake even more terrifying when it is attacking. There is a fabulous myth in India about the origin of these 'spectacles'. One day, Buddha was resting under a tree when he was suddenly attacked by a deadly beast from the forest. A hooded cobra covered the Buddha's head with its large hood and saved his life. In gratitude, Buddha marked all serpents of this species with remarkable spectacles to distinguish them from other snakes.

Cobras are among the most poisonous of snakes. Some 30,000 people die each year in India from the bites of these snakes. It seems very strange, therefore, that these dangerous snakes should achieve such a high level of veneration in India. In some regions, cobras are actually worshipped. In the gardens of Nagpur, and along the Malabar coast, special 'corners' in people's homes are devoted to cobras. The local people believe that the ancestors of the Nagpur rajahs were cobras.

Special festivals in honour of cobras are held in some regions of India. Probably the most famous is a celebration organised in a small village called Shirangala in west-central India. During a cheerful procession, beautifully dressed people carry large pots each containing a live cobra. When the procession reaches the sacred cobra's shrine the snakes are 'begged' by the people carrying them to leave the pots. As the snakes emerge from the pots some of the worshippers catch them by their tails. The alarmed snakes perform their typical display, expanding their menacing hoods and the excited crowds watch while praying. Eventually all the snakes are allowed to go free. Strangely, despite such handling, no one is ever bitten by a snake during these festivals.

In certain regions of India, women who desire to become pregnant place offerings beside the images of sacred cobras. If the woman conceives it is considered to be the direct influence of the sacred cobra. For this reason a happy woman shows her gratitude by erecting a special small shrine in her home.

The human phallus regarded as a symbol of the fertility god Shiva is sometimes depicted in the form of an erect cobra. The snake goddess, Thlen, was once very popular in some parts of India. Rituals related to this god have survived among some people of Assam till recent times. The worship of the snake goddess flourished there as late as 1929 and it was even known for children to be kidnapped and offered as sacrifices to the powerful snake goddess.

Cobras occasionally make their homes inside the tough termite hills of India. When the people who consider cobras sacred notice that this has happened, they perform special rituals including ceremonial walking around the termite hill — because it is now inhabited by a sacred snake it has become a sacred place. People are even known to erect small shrines beside such termite hills.

Tourists who visit India are almost obliged to witness the famous cobra 'dance'. It is an interesting performance during which the snake-charmer shows his remarkable skill. As if on command, the snake rises from the basket to display its hood and starts its famous 'dance' while the charmer plays his flute. Despite the fact that the snakes cannot hear, it is believed that they respond to the music. It is a frightening show since, if the cobra were to strike, it is believed the charmer would be killed immediately. It is also said that in many cases the charmers' profession is not so hazardous as their 'artists' have had their deadly fangs removed.

The snake charmers in Burma, however, are said to perform with truly deadly snakes as their fangs are never removed. The snake charmers have to undergo a long training and their skill in handling the dangerous snakes is based on long observation of the snake's behaviour. This is why some Burmese snake charmers are so brave that they even 'kiss' the deadly snakes and are never bitten during this dramatic show. They seem to know exactly that the snake would not strike at a certain distance or that they are too close for the snake to attack.

The largest cobra is the king cobra (*Ophiophagus hannah*) which lives in India, southern China, Malaysia and the Philippines. It is the longest venomous snake as it can reach 5.7 m. It is also among the deadliest as its venom can kill an adult man in 15 minutes if medical help is not available. The powerful cobra venom is strong enough to kill the biggest land animal, the elephant, and many elephants in India are known to have died from a king cobra's bite. This snake seems to know how to strike to make its bite more efficient. The animal seems to 'realise' that it is difficult for fangs to penetrate the elephant's tough skin, so it usually bites the tip of the elephant's trunk or those areas where the skin is thinnest.

The king cobra is among the most aggressive snakes, and it is known to hunt other snakes. If hungry, it will not hesitate to devour its own weaker mates and hence become a cannibal.

4.
Incredible Sex

Female with a Penis

Even in ancient times, people were interested in the sexual life of beetles. Aristotle noted that it is the female beetle rather than the male, which has a penis-like copulating organ. Aristotle's assertion is not without foundation. In fact there are a few species of beetle whose female inserts a penis-like organ into the male in order to receive the pocket of sperm known as spermatophore. These unique beetles belong to the genus *Cyphon* which is a member of the marsh beetle family (Helodidae). Most species of beetles, however, copulate in the usual way, that is the male inserts his penis into the female.

There are many other peculiarities in the sexual behaviour of beetles. In some species, for example, copulation lasts for many hours, and among soldier beetles (*Rhagonycha fulva*) it may even last for days. There are reports that the copulation of some beetles (e.g. *Golofa pelagon*) found in Bolivia takes place not between the opposite sexes but between males.

In a number of species, such as the cockchafers (*Melolontha* species), copulation takes place while the male is carried by the female who may even feed during copulation.

Sometimes males vigorously fight each other over the waiting female. Stag beetles, which are famous for the antler-like appendages on the males' head are an example. Occasionally, one male starts to copulate with the female while the battle is still going on — very unusual behaviour indeed.

The reproduction cycle of some other beetles is even more amazing. In *Micromalthus debilis* the hatched larva must perform a cannibalistic act in order to develop further. When the larva hatches from the egg, it immediately attaches itself to the mother's skin, inserts its head inside her genital opening and starts devouring her alive from the inside. Once the larva has completed its extraordinary feast, that is, when its mother no longer exists, it moults and attains the pre-pupal and then pupal stages, from which the adult beetle finally emerges.

Although in the majority of beetles the male can easily be distinguished from the female by its larger and more extravagant body, the female is sometimes larger than the male. Because of this disparity, it is sometimes a problem finding a male or a female of a given species.

In 1830 the larva of an unknown beetle species was discovered in Indonesia. It was some 7 cm long and, because of its peculiar shape, it was named 'trilobite larvae'. Surprisingly, no adult form could be discovered for over ninety years. Finally, in 1924, it was discovered that the 'larvae' were actually wingless, adult, female beetles. It was named the net-winged beetle (*Duliticola paradoxa*). It took a long time before the male of the species could be found however, because of a peculiar disparity of size. The male is winged and only about 1 cm long and not only looks very different but is also a dwarf compared to the female. No wonder that the males and females were initially thought to belong to different species.

In some species of beetle, the male's normal 'foreplay' is insufficient to excite his partner and to remedy this he applies special means of attraction. He uses special organs called excitators to make his partner fully aroused before copulation. These organs, which are located on various parts of its body, are normally concealed. During courtship, however, the male displays these organs which are often distinctly coloured. The female becomes attracted to these organs and starts biting them. By doing so she usually becomes sufficiently excited for copulation to take place. If biting one excitatory organ is not sufficient to prepare the female the male may offer another of its excitators. The female's bite also excites the male who begins to copulate more vigorously as a result.

Sexual Champion

The sexual activity of lions (*Panthera leo*) is truly extraordinary. One lion observed in a national park in East Africa mated 157 times in just 55 hours. In the first day and night he mated 74 times with only one female and 12 times with another. This means he mated approximately every 21 minutes and throughout this period he did not eat. During the mating season the lion must fight his rivals in fierce battles for a female. Sometimes, he has to defeat three or four competitors before he can have access to his chosen female. Once the lion has won the battle, the male and female go off together on their 'honeymoon'. They may travel for miles until they find a suitable place where there is sufficient privacy. The 'honeymoon' can last two weeks or even more and during this period they almost totally refrain from eating and hunting.

In ancient times lions were associated with religion. In the ancient Egyptian town Heliopolis there was a lion cult associated with the Sun God Amon Ra. Lions were regarded as sacred and high ranking priests took care of them. They were bathed by the priests in water specially anointed with aromatic oils, they were fed the best morsels and music was played for their pleasure. When a sacred lion died it was embalmed and put into a special tomb with great respect.

In ancient Egypt a war goddess called Sekhmet had a lioness' head. It was believed that she breathed fire against the king's enemies. In Mesopotamia the deity of war, Ishtar, had a lion as her animal attribute.

In ancient Rome lions were exploited for evil purposes. The Romans imported thousands of lions from Africa and used them in cruel spectacles. Christians, dressed in animal skins, were thrown into arenas to meet the hungry beasts. These people were literally torn to pieces by the ferocious animals, to the crowd's amazement. In the Romans' avid desire to eradicate Christians and frighten their opponents, it was said that about 50,000 lions were brought over to Rome during a period of 40 years.

Lions were commonly hunted in ancient times and often emperors and rulers took part in dangerous hunts. One Assyrian monarch shot lions with poisoned arrows from his two-wheeled horse-drawn chariot. A seventeenth century Mogul emperor hunted lions in India with an enormous army of 100,000 men. In some regions of Africa a man who was brave enough to kill a lion with a spear was regarded as a hero. A peculiar method of lion hunting was once popular among the Masai tribes of Kenya and Tanzania. They did not hunt lion for the meat but to show their courage. Their technique was extremely hazardous. They usually worked in a team and when they drove a lion from its lair, they first tried to catch the dangerous animal by the tail. When one of the men succeeded, the rest of the group immediately killed it with their spears. Despite the fact that several men were engaged in this deadly fight, only two of them were praised for bravery, namely, the one who seized the lion's tail and the one who was the first to throw his spear. These men were held in great respect and as one author said: 'No football star on any American campus receives more adulation than did those warriors' who usually wore the lion's mane as a headdress. Until recently, one of the marks of manhood for a Masai boy was the killing of a lion with a spear.

Lions are not hunted in all regions of Africa. In Sudan, the Dinka people never kill lions. They regard them as their ancestors and show them great respect. In fact, they kill other animals and offer them to lions to eat. It is said that these people show no fear of lions and consider them friendly creatures.

Some people take advantage of the lion's ability to kill. The South African Bushmen do not hunt lion, but have a strange method of sharing the lion's prey. When a Bushman spots a lion tearing its prey to pieces, he approaches the beast and asks the lion to leave his prey alone. It is said that, as a result of this peaceful approach, the lion usually leaves the carcass and withdraws, without showing any signs of aggression towards the man.

Lions are not usually very friendly towards men, however. Some even become notorious man-eaters. Lions used to be so abundant in Africa that it was very dangerous to work in some parts of East Africa. When the construction of the railway line from Nairobi to Mombasa was started in 1896, the entire project had to be postponed because of lions attacking people working there. Despite guards, the deadly beasts captured and devoured 28 workers during a short period of time. The man in charge of the railway work in one of the regions became the prey of a man-eater. He had decided, the story says, to kill the lion which was killing his people so he sat at the window of his railway car with a gun, ready to fire. Unfortunately, while waiting for the man-eater to arrive, he must have fallen asleep and the lion must have entered the car and captured his persecutor. Despite the railworkers' efforts their master was never found.

The lion is so powerful that no other animal dares to attack him. Even a huge crocodile cannot win the battle with an adult lion and a huge python can be torn in half by a lion's powerful claws. Some lions do not necessarily hunt for food. A lion killed a hundred pigs in one African village in one night, evidently for the pleasure of killing.

Interestingly, it is not usually the male who hunts but the female who is responsible for feeding her family. The male usually arrives after the female's successful kill and he eats the best parts of the quarry first, while the real hunter patiently waits her turn.

Love Darts

Some snails exhibit truly incredible sexual behaviour. Two snails approach each other and during a kind of courtship they exchange the symbols of love by firing 'love' darts at each other. These love darts may penetrate the flesh of the other snail to a considerable depth, but they are eagerly accepted because they are powerful sexual stimulants. Love darts are usually about 2 cm long — quite long compared to the small size of the snail. Depending upon the species, they may be multipointed, spiralled or may even resemble a dagger. To make such a love dart the snails have to gather limestone, because calcium carbonate is the main component of this unusual weapon of love.

This 'love affair' does not take place between a male and a female. These snails have only one gamete-producing organ known as an ovotestis. Initially the ovotestis only produces sperm, so the snails can be said to be males only. After the love affair, described above, mutual insemination takes place and the sperm are stored in a chamber leading to an oviduct which is scientifically referred to as a spermatheca. After a period of time, the ovotestis produces eggs. At this stage, therefore, it can be said that each snail has become a female. The sperm, which has been stored since mating, then fertilises the eggs produced. Some snails, however, are unisexual.

Perhaps the strangest sexual behaviour can be observed among certain close relatives of snails, namely slugs of the genus *Agriolimax*. When two slugs meet they attempt to gnaw off each other's penises, a kind of mutual castration which is hard to imagine. The castrated slug's penis does not regrow and, because of this unusual operation, the slug now behaves as a female and is forced to produce eggs.

Snails adapt easily to harsh conditions. Some can survive the driest climates, while other creatures perish.

The snail can enter a state of so-called suspended animation during which it appears to be dead. It does not eat or move and in this state of lethargy it can survive alive for months or even years. Incredibly, one specimen of the *Oxystyla copax* snail reportedly survived in suspended animation for 23 years. The snail then started living a normal life as if nothing had happened.

Although snails are usually regarded as rather weak animals, they are in fact remarkably strong. The Roman snail (*Helix pomatia*) is so strong that it can pull objects 200 times its own weight and can carry a load on its back 12 times its own weight.

Snails are most famous for their lack of speed. They are in fact one of the slowest moving creatures in the world. They usually move on land at a rate of some 2.5 cm per minute and therefore take 25 days to travel one kilometre. Because they are so slow they carry their 'house' wherever they go.

Snails have an incredible number of tiny teeth which are arranged in several rows. In the floor of the mouth of one snail species some 27,000 teeth have been counted and they can be exceedingly sharp. A snail can easily cut its way out of a cardboard box and meat-eating snails are even known to use their sharp teeth to cut through the tough shells of other snails to reach the flesh.

The first shell collectors were not humans but a particular species of snail. These snails collect pieces of broken shells left by other animals, or bits of coral and attach them to their own shells. As a result, the snail looks like a 'little pile of broken shells', which is an excellent camouflage against predators. These collector snails live in tropical seas.

The world's largest land snails are the *Achatina* species which are often some 30 cm in length and weigh 200 grams. The world's largest specimen found in Africa was 30cm long and weighed 900 grams. These snails occur in various parts of Africa and in the Pacific Islands. They are real giants and in some regions people eat their 'meat'. In Ghana, for example, these giant snails have traditionally been a source of animal protein and people spend much time collecting them.

Snails have been used to feed fighting troops. When the Japanese conquered some of the Pacific Islands, they fed their troops on snails which could be gathered in the forest. The American soldiers, who later arrived in these regions, must have accidentally taken these giant snails home, since they were later discovered in California where they multiplied at an amazing rate — they can produce 11 million offspring in just five years. As a result, the snails became a real pest in some parts of America. Poisoning and burning had to be done on a large scale to keep the snail population under control.

The snail was regarded by the Aztecs of ancient Mexico as a symbol of fertility, successful conception, pregnancy and childbirth. In northern Africa snails were used as a cure for various ailments, especially gout. The shells were removed and pounded and then applied to the affected part of the body to alleviate pain. In some regions, powdered shell was used both as medicine and as a love potion. In some societies small, pretty snail shells were used as good luck charms and even as money.

Bizarre Lover

For deep-sea fish which live on the pitch-dark sea-floor it is not easy to find a sexual partner. It is even more difficult when the sexes differ in size. A classic example is the *Ceratias holboelli* fish, the male of which is so small that it was thought to be a different species. While the male is only some 6 cm long, the female of the same species is a real giant as it can measure some 1.2 metres. Taking into account the weight of the two partners, the difference is even more pronounced as some females may weigh 1,000 times more than the tiny male.

It is the tiny male who makes the main effort to find a sexual partner while the female only facilitates this task by emitting a distinct scent to announce her presence. The search for the opposite sex is essential for the male since at a certain stage of his development he loses the ability to feed and dies if his search is unsuccessful. Using his highly sensitive odour-tracking sense, the male locates his partner in the immense sea areas often far from his usual location. According to some observers the male of the species is among the most odour-sensitive animals in the world.

He does not try to copulate as soon as he finds his sexual partner as at this stage his gonads are undeveloped. The tiny male attaches himself to the huge female by making incisions in her skin with his sharp teeth anywhere on her body, including her head and not necessarily near her genital orifice. Finding his sexual partner is so important for the male that he remains with his 'wife' for ever. The tiny male gradually grows into her body and loses almost all his internal organs. He only retains his respiratory system to have as a source of oxygen to keep living. He does not need to eat as nutrients are supplied from the bloodstream of his female partner. So the male becomes a virtual parasite of his wife. He loses his eyes, mouth and digestive tract, but develops his sexual organs. In fact, the male becomes an ever-ready sperm bank for his 'lady', but only releases sperm to fertilise eggs upon her command, which she gives through special hormones secreted in her blood. So the tiny male is important for the female, while his fate depends on her life. When the female partner dies the tiny male must also die as he has lost his only source of food.

The huge female of this species may have more than one 'husband' attached. A large fish with several males attached in various parts of her body looks very strange to anyone who has the chance to observe such a creature.

Playboy

Some birds found in Australia and New Guinea have courtship rituals which are probably the most extraordinary in the whole avian world. To seduce their partners these birds, called bowerbirds, apply an unusual method. The male builds an elaborate pavilion called a bower which is not a nest, but a kind of structure built for the sole purpose of attracting females. These bowers, the structure of which varies according to the species of bowerbird, are made of grass and twigs and may be up to 3 m high. When explorers first saw these bowers they thought they were toys made to amuse children.

It is not the pavilion itself which attracts the female, but the numerous objects such as colourful feathers, shells, flowers or berries which decorate the arena or entrance to the pavilion. The male bird works hard to make the bower as attractive as possible and if humans live nearby he may collect buttons, glass beads, coins and even such objects as toothbrushes or broken spectacles. Birds have even stolen keys from cars. In one case a glass eye disappeared from a glass when the owner fell asleep in his room. Precious jewellery may also attract the bird's attention. A precious sapphire disappeared once and the owner accused his servant of stealing it. To his surprise the ring was later found decorating a bower in the forest nearby.

A bowerbird may use an enormous amount of plant material to construct his bower. A Lauterbachs bowerbird (*Chlamydera lauterbachi*) of New Guinea may use 3,000 wooden sticks, 1,000 stems of grass and collect many kilograms of small stones to decorate the stage of his bower.

Some bowerbirds show a rather strange preference for the type of items they use to decorate their bowers. The spotted bowerbird (*Chlamydera maculata*) mainly collects small, bleached bones of small animals. It was estimated that one such bird gathered 1,300 small bones in his bower. Because of this strange habit this bird became known as a 'sepulchre bird'.

Each species has a preference for a different colour, for example, blue or red. The satin bowerbird (*Ptilonorhynchus violaceus*) found in Australia has a strong preference for blue objects. This bird mostly collects blue beetles, feathers or flowers and picks up blue objects left by people, such as buttons or glass from broken blue bottles. This bowerbird is known to go into laundries through open windows or doors and pick up bluish coloured items. He may also take blue pegs from washing on the line.

This clever artist can even paint his pavilion blue. He collects blue berries and, after crushing them, uses the fruit juice as paint. Holding a fibrous 'brush' in his beak, he behaves like a human painter. He soaks the 'brush' in the paint and then paints his bower. He constantly checks his work and when the paint is washed away by rain he repaints it.

The laborious efforts of the male have only one purpose, to impress the female. The female bowerbird is only attracted to the specific colour used by the male of her own species and in this way there is no mixing of species. The male seems to realise that he will only win the competition between the many males of his species by constructing the most beautiful bower.

When the decorations are in place and the breeding period starts, the male stands on the stage and waits for a female. The arrival of a female triggers an intense display. The male starts to 'dance' and call, often imitating human, cat or other bird calls. If the female is suficiently impressed with the display and the elaborate pavilion stage she enters and ultimately copulation takes place in the bower.

After mating, the male completely loses interest in his lover who flies away to continue her own life outside the pavilion. The male is a promiscuous bird and starts looking for another sexual partner. The bowerbird is so obsessed with love-making that he tries to invite as many females as possible to his bower. No wonder that some biologists have called the bird a 'playboy'.

The female seems to accept the lack of a strong relationship with her sexual partner, as she flies away soon after mating and builds her nest without any further involvement.

It was thought that constructing bowers was an inborn instinct of these birds, but recent studies have revealed that it is not innate, but rather a learned behaviour. The young male starts by building a very poor pavilion, but gradually gains experience and improves his skills.

Lovelights

When the first British explorers arrived at night in the Blue Mountains on the island of Jamaica they thought that a crowd of natives had come to greet them with thousands of candles. They soon learned, however, that they were fooled by a multitude of insects glistening in the darkness. The luminescent insects are in fact a certain species of beetle which are commonly called fireflies.

The luminous beetles occur in various regions of the world but are especially common in warmer areas. Central American Indians catch the click beetles (family Elateridae), which they call 'coccoji', and, by threading some on a string round their necks have an excellent source of light when walking in the forest at night. The beetles shine so brightly that they can be used as candles inside a house.

Native girls in some parts of Brazil tie the luminous beetles in small gauze bags to adorn their hair and at night they look very attractive. During the Firefly Festival near Kyoto in Japan the luminous beetles kept in small cages are used by the local girls in their boats at night. The girls boast that these natural lamps are stronger than the light of the stars.

The light emitted by certain beetles is so strong that in China, poor students used them to read by at night. Strangely, fireflies once saved a life.

In 1898 the Americans were fighting in Cuba and once, while a surgical operation was being performed in a field tent the lamps suddenly went out. Fortunately someone quickly brought a bottle full of shining beetles and the surgeon was able to complete the operation under this natural lantern.

During the Second World War, insects helped American troops who were fighting in France. Using headlights on motorcars was strictly forbidden, even when driving down the narrow and winding roads in Normandy. To the drivers' surprise, however, the dangerous roads were lit by thousands of miraculous minute lights which suddenly started to shine.

The most fantastic 'fireworks' are produced by luminous beetles of the genus *Colophotia* in South-east Asia. In the evening the mangrove trees along the river-banks become illuminated by millions of tiny lights. The light is so strong that every tree or boat on the river is seen as clearly as if it were day. But the light is not constant and only lasts a short while. The millions of beetles sitting on the trees, without exception, switch their lamps on and off at exactly the same time as if responding to a mysterious command. What is more amazing is the insects blink at an exact rate of two blinks a second.

This fantastic spectacle is only performed by the male. The females of this species cannot emit light in response to this evident love call but they know where to go. In some species the females are unable to fly and so respond to their males' love signals by using their own light flashes. Guided in such a way, the males can easily find their love partners hidden in the grass.

Each species of luminous beetle has a specific blinking pattern which is only recognised by the opposite sex of its own species. This is very important as there are so many beetle species living in the same area. Although this communication system works extremely well there is still room for abuse. Certain luminous beetles, for instance, the female *Photuris versicolor* use the light code of other beetles to their own advantage. They can mimic the light signals of other fireflies perfectly. They give such accurate responses to the males' love signals that the males believe they are their own females responding. When the unsuspecting males come to make love, they are attacked by the 'foreign ladies' who are not interested in them as lovers, but quickly overpower and devour them.

There is another disadvantage to the fireflies' remarkable luminescence, as their signals break their camouflage for predators. Light-emitting insects are easy prey for various frogs, some of which are known to consume so many during their mating season that they themselves start to shine in the darkness.

The fireflies' light-emitting organs are located in various parts of the body. The click-beetle, found in Central America has luminous glands on the front of its body which emit a strong white light with a third red light at its rear which is rather weak. The two strong front lights are likened to the headlights of a motorcar while the rear light looks like the one which illuminates a car's registration number. This beetle is often called in

America the 'ford bug'.

How certain beetles are able to emit light remained a mystery for many centuries. We now know that it is the result of a complicated chemical reaction, the main substance being luciferin. This substance emits light when affected by a special enzyme and in the presence of oxygen, which must be supplied by the insect's tissues. It is a typical cold light and a firefly is an extremely efficient light producer as it converts some 98 per cent of energy used into light while an electric bulb converts only 3 per cent.

Sexual Lunacy

It is hard to believe, but the sexual life of a sea-worm (*Lisidice oele*), known as 'nyale', is governed by the moon.

Precisely after the full moon of February and March, the mysterious reproduction takes place and an incredible mass of worms appears on the sea's surface. Islanders living on Sumba, between Bali and Timor, carefully observe the moon phases in order not to miss the incredible harvest. These worms suddenly appear in such a mass on the sea surface that people rush down with their canoes to gather them with baskets. This harvest is regarded as a miraculous gift of nature.

It is a highlight of the year and the fantastic appearance of worms has been explained in various tales. According to one, there was once a severe famine on the island and the ruling king, Rasa Wulan, was asked to do something by his people. His priests advised him that the only way to please the gods responsible for their calamity was to make a human sacrifice. The king decided to offer his most beautiful daughter, Nyale, to the gods. The girl was killed and ritually dismembered and all her body parts were cast into the sea. The gods were so appeased by this great sacrifice that they magically transformed Nyale's body into various sea creatures, which could be eaten by the natives. A shark was created from the tibia, her liver became a squid, her fingers were changed into an octopus and crabs were made from her feet, Her beautiful hair, lips and blood became the marine worms called 'nyale worms'. Since then, the story says, the miraculous worms return each year to remind the islanders of the heroic martyrdom of the king's daughter.

Similar sea-worm harvests also take place in other regions. In Samoa and Fiji, masses of Pacific palolo worms (*Eunice viridis*) also appear twice a year, in October and November. But people are only interested in the big palolo 'crop' which occurs one week after the November full moon. People rush down to collect the worms, which appear in enormous numbers on the sea's surface. The water looks like a huge vermicelli soup. People gather them and some eat them immediately with evident satisfaction.

In the past, gathering the mysterious worms was under the strict control of local shamans. They were the first to arrive at the sea and the first to collect the delicious worms. The excited crowds had to wait until the shamans gave a signal to join them in the harvest. It was believed that if someone started collecting the worms before the shaman, he would be severely bitten by dog-spirits, which would make the 'offender' insane.

By observing the moon phases, the worm harvest could always be predicted with such precision that it used to mark the beginning of the Samoan New Year.

The influence of the moon on the worms' sexual reproduction has always fascinated biologists. The mysterious palolo worms, which are some 40 cm long, live in burrows among coral reefs on the sea bottom.

These worms have highly segmented bodies and each worm is composed of up to one thousand segments. As if on the moon's command, millions of worms suddenly and simultaneously break in half at the sea bottom. The sexual halves then wriggle in an effort to reach the surface quickly. Here they discharge eggs and sperm which causes their complete disintegration.

This remarkable discharge takes place at dawn, just before sunrise. The islanders are aware of the strange habits of these worms and they come at dawn, just before the worms start discharging eggs and sperm.

The worms' non-sexual halves, which remained in burrows at the sea bottom, soon regenerate the lost sexual part and the whole mysterious cycle of reproduction is repeated at the right moon phase the following year.

Grotesque Lover

With its enormous, thick tentacles around a distinct head with large bulging eyes, the octopus frightened fishermen and became known as the 'devil fish' of the ocean. These sea monsters were believed to capsize boats and devour fishermen. Octopuses became the subject of many myths among the natives of Pacific regions. There is a belief in the Kiribati islands that the entire archipelago was created by a giant octopus-god who used his powerful arms to gather stones from which the islands were built.

Despite the respect given to these animals the inhabitants of Kiribati now hunt octopus which is regarded as an excellent source of food. In some Kiribati islands a rather strange hunting technique is still practised. The fishermen work in pairs. When they spot a large octopus, one of the divers directly approaches the animal to attract its attention while another man immediately comes from the other side and bites the octopus between its bulging eyes. Such a bite is deadly to even a huge octopus, but should the man miss his target the octopus will fiercely defend itself and strangle the man, who may drown as a result. The Kiribati hunters know well that the huge animal's most vulnerable spot is its brain, located between its eyes, and once it is destroyed by the man's bite the octopus loses its battle.

These animals are very intelligent and often find a successful method of escape, even from a huge attacking shark. It has been reported that even when mauled by huge jaws it is able to rescue itself by quickly moving onto the fish's back where it cannot be reached.

When looking at the large octopus it is difficult to imagine that such a grotesque animal could be desirable. It is, nevertheless, very attractive to members of the opposite sex.

For the purpose of reproduction the male octopus has a specialised arm, which is usually longer than the others and is neither a tentacle nor a penis. On the underside a groove runs the length of the arm to the tip, which is shaped like a spoon.

When a pair of lovers meet, they perform a kind of courting ceremony during which the male massages the female's body with its reproductive arm. After a while, both partners are sexually aroused and this is signalled by the sudden colour change of their bodies. Their strange copulation then commences. The male inserts the spoon-like tip of his sexual arm into the female's mantle cavity and passes bundles of sperm along the groove so that they emerge just at the mouth of the female's egg duct.

During this peculiar copulation, which may last an hour, the male is sometimes too vigorous and his sexual arm breaks off and remains in the female's mantle cavity. The females of some octopus species can have several of these detached 'lover's arms' in her body and until the nineteenth century these arms were thought to be parasitic worms. Sometimes these sexual arms are shed by the female and they can be seen floating in the sea as if looking for a sexual partner. The 'wounded' male's reproductive arm eventually regenerates.

Although female octopus deliver some 200,000 offspring, only one or two of them reach maturity.

Octopuses used to be much larger in the past and some were real giants of the sea. The discovery of an incredibly large octopus once created a great sensation. In November 1896 the remains of a huge marine animal were found on a beach in South Florida, USA. Although they only had part of an animal, after careful examination it was concluded that it must be part of a huge octopus with an arm span of some 30 m and weighing about seven tonnes. This unusual octopus was named *Octopus giganteus.*

Soon after the discovery, however, another scientist rejected this identification and claimed that it was part of a sperm whale. There was therefore great controversy over the discovered carcass and after a few years the whole story was forgotten. Seventy-four years later, a scientist started to analyse tissue samples of a huge octopus-like creature which were preserved at the University of New York. After performing biochemical tests, he came to the conclusion that the sea monster discovered was actually a giant octopus. Not all marine biologists agree with these findings however. They claim that the world's largest octupus was found in 1957 off Canada. This giant had an arm span of 9.6 m.

The most common, however, are not the giant octopuses but rather smaller animals with an average 2.5 m arm span. The smallest known is *Octopus arborescens*, which is found in Sri Lanka and whose radial span is only 5 cm.

These creatures have a remarkable ability to change the shape of their bodies to such an extent that they can squeeze through very small cracks. Some octopuses which were kept in the New York Aquarium were said to cause problems because they could escape through a very narrow opening. An octopus which measured some 90 cm across was able to squeeze through a slit in a box which was only 0.3 cm.

Octopuses like to get into almost any empty container they can find on the sea bottom. They will make homes of empty tyres or bottles — even an empty human skull. A sunken ship is a favourite place for octopuses where they often make strange, underwater cities.

Surprisingly, these peculiar animals can become pets and when kept in an aquarium they recognise their masters and are said to eat from people's hands. They are quite intelligent animals and can be trained to remove a cork from a bottle. They can differentiate between sizes and shapes of various objects and remember things for quite a long time. The octopus is believed to be the most intelligent invertebrate and is said to have an intelligence level similar to that of a domestic cat.

5.
Horror of Love

Saintly Killer

The most formidable of all insects are mantids, also known as mantises. Some of these insects are famous for their peculiar position when resting. Its forelegs are held together and raised as if in prayer. No wonder that when Linnaeus examined the strange behaviour of one European mantid he named it the praying mantis (*Mantis religiosa*).

According to Muslim tradition this insect always prays with its face turned towards Mecca. As one ancient author reported: 'It always holds its forefeet like hands, praying, as it were, after the manner of the Divines who in that gesture did pour out their supplications to the God'.

Mantids were attributed with mysterious powers to foresee the future and predict disasters such as famine or death. The Hottentots of South Africa believe that mantids are the incarnation of their tutelary divinity and they are regarded as sacred in some regions.

The insect's saintly attitude, however, is actually a pose to fool its prey and it is in fact a ruthless killer. In its praying posture the mantis is only waiting for a victim to come closer so that its grasping legs can strike out and seize the unsuspecting prey with lightning speed. Held in a pincer-like grip, the victim is brought to the mantid's jaws and devoured.

The mantid will attack and kill creatures much larger than itself. They are known to kill small frogs and lizards and occasionally they even attack their own weaker mates. Larger mantids have even been known to attack and devour young birds, capturing them in their nests. B. S. Morse witnessed such an event in Australia in 1922 when a tiny yellow-rumped finch was killed by a large mantid. The ferocious insect killed this tiny bird and extracted its brain from a hole made in its head. Some biologists have called these insects bush terrorists.

The most unusual feature of these insects is the female's attitude towards the male when he approaches to mate. In fact this is one of the most dangerous moments in the male's life as the 'loving' female has an unusual appetite and usually devours her lover. To minimise the risk of being eaten, the male, who is usually much smaller than the female, approaches her very cautiously from behind and leaps on her back. But the male can only occasionally position himself so that the hungry female cannot reach him with her grasping forelegs. In other words, the male only very rarely escapes after copulation. Usually he becomes a meal for the female after mating. In fact, even during copulation, the female may start her feast by biting off the male's head. Amazingly, this does not end the copulation, which may take hours, as the beheaded male intensifies its copulatory movements until its whole body is devoured. Biologists were puzzled by the beheaded male's behaviour as it was difficult to explain why it would mate more effectively when it is partly consumed and deprived of its head. It has been revealed that by removing her mate's head, the female releases the control of the male's brain over its sexual organ which leads to the incessant copulatory movement. A truly amazing 'love story'.

Although there is no advantage to a male mantid in being eaten by his lover, there is some advantage for the female who gets a nutritious meal and becomes stronger and more fit for laying her eggs. Although female mantids are famous for their cannibalistic behaviour, there is an American species of mantis whose females behave 'decently'. It is not that they are merciful, however, but simply that there are no males in this species. The offspring is produced from unfertilised eggs in a process known as parthenogenesis.

Mantises became totems of the Asmat tribe of New Guinea, and one of the probable reasons for such a choice is that both the tribesmen and mantids were notorious cannibals.

The unusual ability to fight and to devour its own mates used to provide amusement for some people. In China there is a large mantis which is some 10 cm long. A pair of these mantids would be placed in a bamboo cage and the crowd would watch the insects fight. The highlight of the show was the cannibalistic feast of the victorious insect.

The Japanese were also impressed with the incredible fighting capacity of mantids and so these insects became a symbol of bravery. The Japanese call mantids 'kamakiri', meaning sickle cutter, to emphasise the insect's skill. Even the design of certain samurai swords was inspired by mantids and mantids remained a symbol of unrelenting fighting.

These strange insects were also believed to have a miraculous healing power. As early as the Confucian era in the fifth century BC, mantid eggs were boiled in water and the potion was used for healing severe battle wounds.

In some regions, however, mantises are regarded as bad omens. Some people believe that a person becomes ill because a mantid has looked upon him with its bulging eyes. In Sardinia, people avoid touching mantids in the belief that they will bring them misfortune.

Surprisingly, these ferocious mantids, which are so aggressive towards many animals, can be tamed and become pets. Tamed mantids are said to be able to learn how to drink out of a teaspoon and eat chopped meat from a saucer. As one mantid 'trainer' proudly claimed: 'From his corner, the mantid will follow your every move, his head cocked perkily to one side, like a playful puppy.'

Mantids do have enemies and they often try to avoid being detected by the use of camouflage. Some mantids resemble leaves and may not only be green but can even be flattened and veined to resemble a leaf. The most famous is the camouflage of the Malaysian mantid *Hymenopus bicornis*. It has pink and petal-like appendages on its body and legs which make it difficult to distinguish from the orchid flower because it imitates the flower so well. With this excellent disguise the mantid is usually safe from predators. In addition this provides the insect with the opportunity to catch those busy insects which arrive to pollinate this unusual 'flower'.

Bizarre Love

Many species of spiders are among the most unfortunate of lovers. The mating procedure for the male of many spider species is extremely risky. Many female spiders are larger and much stronger than the male and they are always so hungry that soon after copulation they devour their lovers on the spot. In fact, so many males lose their lives as a result of their sexual partners' cannibalistic habits that it is hard to find a male of some species. Some species are even called widows — for example the *Latrodectus mactans*, which is known as the black widow.

To stay alive after copulation, some spiders use special tricks which are sometimes effective and save their lives. In some species the clever male brings an insect as a gift for his partner before copulation, usually larger and more appetising than the male himself. After copulation, the happy, but hungry, female turns to the larger insect and eats it while the male quickly retreats, saving his life.

In some species, in order to save his life the male brings his gift wrapped in silky threads making a kind of cocoon containing a juicy insect. While the hungry female unwraps the prey, the male has enough time to perform his act of love and escape unhurt.

There are also crooks among spiders who behave in a very peculiar way. They catch a large juicy insect, suck out all its juice, and then wrap it up to make a nice cocoon. This gift is offered to the female at the beginning of the love affair. When the 'happy' female unwraps the gift she finds a useless husk inside, but by that time the clever male has completed his sexual act and escaped unhurt.

Amazing sexual strategy is employed by the crab spider (*Xysticus cristatus*). The male of this species immobilises his female by tying her down with a silken web. Once she is tied up, the male then creeps under her body and transfers his sperm into her genital opening. Thus the hazard of being attacked by a female after copulation is cleverly avoided by the male's invention of this form of 'safe sex'.

Interestingly, some male orb weaver spiders avoid entering the female's web during courtship. He cuts several threads of his web and lays down his own strand, called the 'mating thread', from which he attracts his female with a special dance. The female climbs out onto the thread and mating takes place. But when the female threatens to eat her 'lover' he simply cuts the thread and swings away unhurt. By this unusual courtship technique, the male avoids becoming his partner's prey after mating.

Spiders may copulate for a few seconds or for many hours, depending upon the species.

In some species, the male does not need to be afraid of the female's deadly bite. The very large female *Nephila* spider can be some 1000 times larger than the male. It is difficult to imagine a sexual relationship between partners of such very different sizes, but they do take place. One could expect that the female would devour her partner immediately after copulation, but surprisingly this does not happen and the male escapes unhurt. It is not that she is merciful and prefers him to remain alive; she simply ignores him, she does not consider such a minute spider to be a worthwhile meal.

Some male spiders, for example the orb spider (*Araneus pallidus*), do not even try to escape after copulation. In fact, he is unable to perform his act if the female does not assist him. She catches the male's abdomen and so helps the male insert his palps into her sexual opening to transfer his semen. If the female did not do this, the tiny partner would immediately slip off her huge body and be lost. Unfortunately for the male this ends tragically as, immediately after copulation, the female devours her mate. In this strange relationship, the male's unhappy fate is a 'technical' necessity. He must actually sacrifice his life for his species to reproduce and survive.

Male spiders have no penis but despite this formidable handicap they try to copulate in the normal way. The male spider's sexual procedure is unique as he copulates using a pair of palps or feelers, which are designed to be filled with semen exuded from pores on his abdomen. So the spider prepares for the mating by dipping the ends of his feelers into the ejaculated fluid and after a kind of courting the male inserts them into the female, performing the strange act of love.

6.
Peculiar Aphrodisiacs

Magic Horn

Rhinoceroses are among the most extraordinary of mammals. To many people these animals, with their massive size, bare skin and grotesque general appearance, look like extinct dinosaurs. Most impressive of all are the horns or horn on their snouts. Depending upon the species, they have one or two horns, and their size makes them a formidable weapon. The longest one known measured 108 cm, although in the past some animals had horns 2 m long.

In the Orient these phallus-shaped horns attracted people's attention in the past. Impressed with the horn's similarity to the human sexual organ, people started using powdered horn to enhance men's sexual performance. It became a famous aphrodisiac. People must have also noticed the rhinoceroses' own unusual sexual powers. This bulky and unattractive animal is among the most vigorous lovers. It usually copulates for up to 60 minutes, which is really impressive, since most mammals copulate for only a few seconds. In one case a rhino was seen copulating for over 80 minutes, but this is rather exceptional. The Indian rhino has been seen to ejaculate every three minutes.

Little wonder that rhinoceros horns have become one of the most expensive aphrodisiacs and thousands of these remarkable animals have been slaughtered to supply the growing demand for horn. One kg of powdered horn may fetch the astounding price of up to $70,000 on the black market.

Despite the fact that modern science has been unable to discover any chemical ingredients in rhino horn which could enhance sexual power, the belief in this famous aphrodisiac still persists and in many pharmacies in Asia the powder is sold as a sexual stimulant and the illegal trade still flourishes.

The belief in the miraculous power of rhino horn is so strong in some Arab countries that it has become a symbol of masculinity. In

Northern Yemen, an adult Moslem has to wear a ceremonial dagger called 'jambia', which he keeps in his leather belt. The most prestigious is a dagger which has a handle carved from the horn of the black rhinoceros (*Diceros bicornis*). For this reason there is a great demand for rhino horn in that country. It has been estimated that some 8,000 rhinos were killed between 1969 and 1977 for the Yemeni market.

As it is risky to approach a rhino in the wild, it has become a test of bravery among young Masai boys in East Africa. They have a peculiar game. When the boys see a rhinoceros resting somewhere in the savannah one of the boys approaches the beast very carefully from behind and places a stone on its back. He must do it in such as way that the beast does not notice. Once the stone is laid another boy must remove it. The game goes on until the rhino wakes up and starts to move.

As well as rhino horn, it is believed, in some regions, that the penis has a similar miraculous power to enhance men's sexual performance. The men in some regions of Borneo even try to enhance their sexual power by adapting their penises to resemble those of the Sumatran rhino. They have noticed that the rhino's penis has two strange sausage-like flanges about 5 cm long, which dangle from the penis shaft like crossbars on both sides of the sexual organ. The men insert pieces of bamboo into their penises to resemble those of the sexually potent rhinos.

Even rhino's blood was believed to have powerful qualities. There used to be a strange ritual in East Africa involving the use of rhino's blood. After slaughtering a big rhino people used to 'bathe' in its blood in an attempt to gain its great strength. Captain C. R. S. Pitman, who witnessed this event in 1920, was astonished by what he saw. He reported: 'Having stripped themselves of their scanty body adornments, these . . . savages, with evident delight, smeared themselves from head to toe with rhino's blood.'

There was also a strange ritual in India related to the Indian rhinoceros (*Rhinoceros unicornis*). A freshly slaughtered beast was gutted and several people would squat inside the giant animal's body reciting prayers to their gods. The rhino's blood was carefully collected and sacrificed to the gods.

In some parts of India, it is the animal's urine which is considered to be the most useful medicine. Demand for this strange cure is so great that some zoos sell their rhinos' urine to local people. Sometimes people queue for hours to collect the precious liquid to cure their ailments. The customers usually work in pairs. When the rhino starts to urinate, one man jumps in front with bananas to keep the animal busy, while the other approaches it with an empty container.

This strange looking animal was first described by Marco Polo during his travels to Sumatra in 1298. Despite this discovery, the rhinoceros remained unknown to the western world for several centuries. The first live rhino in Europe was the gift of an Indian king to the Portuguese king, Emanuel, who, in turn, decided to send the peculiar animal to the Pope. Unfortunately, the ship, with its precious cargo, sank before it reached Italy. Although all the people and the rhino perished, the famous animal's carcass was later found on a beach. It was skinned and stuffed and looked so good that it was still presented to the Pope.

When the first live rhinoceros was brought to Paris it was exhibited in a special wheeled cage which was drawn by eight horses, causing a great sensation among the crowds.

These powerful animals were even used on the battlefield. In India, special iron tridents were attached to rhinos' horns, and when the enemy troops saw them they panicked, losing the battle before it had started.

Unfortunately, because this animal has traditionally been used as an aphrodisiac and cure-all, it is among the most persecuted animals in the world. Despite the fact that most countries have agreed to the ban the trade of rhino-horn, poachers still kill them for illegal export. Conservationists fear that if this illegal trade persists, four of the five known rhinoceros species will soon become extinct.

Mysterious Tears

Early European sailors brought back incredible stories about a strange creature in the Red Sea which was half woman, half fish. They claimed that they had seen this creature nursing its offspring in an upright position, like a human mother nursing her baby. There was some truth in this story since there are some sea mammals which have two nipples on their chests and a female may occasionally hold its baby in an upright position with one of its flippers.

This strange animal, the dugong (*Dugong dugon*), is a huge bulky creature with a strange face, hence another of its popular names, 'sea cow'. It remains a mystery how this mammal could have been 'transformed' into a beautiful sea-maiden, or mermaid. It is believed, however, that it was this animal which was the inspiration for Homer's sirens who tempted Ulysses in *The Odyssey.* Only people with a vivid imagination could believe that a female dugong was a beautiful sea-maiden.

Some people, however, took advantage of this siren story and there was to be a factory in Japan which produced artificial sirens for which there was a big demand. Until recently, in Aden in South Yemen, stuffed dugongs were shown to tourists as real mermaids. It used to be believed that these strange sirens could speak, but in a language nobody could understand.

Dugongs are sea mammals which spend most of their time in the water as they are vegetarians and feed on seaweed. The dugong is a very bulky animal which can be some 3 m long and weigh up to 400 kg. Dugongs live in warmer coastal waters from Australia to Madagascar and used to be among the most common sea animals. In fact some early European settlers once saw an unusual horde of dugongs in the sea off northern Australia, which covered an area 4 km long and some 270 m wide.

These strange looking animals have one very 'human' characteristic. When the animal is upset, especially in captivity, it 'cries' tears, as if showing its unhappiness. In a zoo in Sri Lanka an 'imprisoned' dugong was said to have wept for several months. The strange 'tears' are not the same as those shed by humans, however, but are actually an oily secretion, produced by special glands, which prevents eye irritation in salty sea water.

The 'tears' shed by a supposedly unhappy 'mermaid' have a special significance to people in some regions. Local fishermen in Madagascar used to purposely molest captured dugongs so that they would produce lots of 'tears'. Young dugongs are easily caught when their mothers have been killed, and in Indonesia these babies are often imprisoned in order to have a steady supply of tears from the crying animals. Dugong tears are a popular love charm. These charms, known as 'air dugong' were very popular among the men. They carried a swab of wool soaked in the mysterious 'tears' and believed that they would bring them luck with women. Dugong teeth were also regarded as having mysterious powers. Made into cigarette holders, it was believed that smoking tobacco through such pipes would give men enormous sexual drive. Dried dugong penises were also commonly used by men as an aphrodisiac.

Peculiar customs can be observed among Madagascan fishermen who hunt dugongs. A man who kills a dugong cannot sell the meat on local markets unless he performs special rituals which involve the fisherman's oath that he did not secretly try to have unnatural relations with the 'fish-woman'. Strangely, during initiation ceremonies in some coastal regions of Papua New Guinea a female dugong represents the human female.

In some regions of Nigeria the dugong is regarded as sacred and believed to be the incarnation of the souls of human ancestors. This, however, does not stop dugongs being hunted for food. But when the sacred animal is killed a special cleansing ceremony has to be performed by the one who killed the animal. The hunter must not leave his home for three days. His body, is rubbed from head to toe with a yellow dye by the women of his family. On the last day of this purification ritual, the dugong's meat is offered to each member of the community. Finally a small portion of the animal's flesh is placed in a shrine as an offering to their ancestors.

The bulky bodies of dugongs contain a large amount of oil, which Australian Aborigines have known for a long time. They used the oil as a medicine which was believed to cure any disease. The belief in the power of this oil was so strong that, in the past, tribal members were carried from the interior of the continent to the seaside so that they could be treated with the miraculous dugong oil. When the first European settlers learned about the medicinal value of this oil they also started using it. It became a famous cure for arthritis in the old days. To supply the market, special fisheries were established in Australia and many 'beautiful' sea maidens were slaughtered.

The extensive hunting of dugongs almost led to the extinction of this once common animal. Fortunately, dugongs are now protected in many regions, including Australia. It has such a great significance among certain Aborigines in Australia, however, that special permission to kill a limited number of these animals is occasionally given by the authorities. Before the Christmas holidays, a special permit to hunt dugongs is given to the Aborigines in Arnhem Land and the Gulf of Carpentaria, so that they can have their favourite meal during the traditional festivals.

Sex Invigorating Delicacy

Hunting birds for meat is quite a common human activity, but hunting for the birds' nests appears to be a figment of the imagination. Yet in some regions harvesting such nests has become a very lucrative business.

The birds which build these precious nests are the edible nest swiftlets (*Aerodramus fuciphagus*) which live in dark caves along the coasts of Southeast Asia. As many as one million birds can live in some caves in total darkness. These birds, which weigh no more than 10 g, produce rather unusual nests which are made almost totally of their saliva. These translucent and rubbery nests have been a famous Chinese delicacy and aphrodisiac for centuries.

The origin of the strange habit of eating birds' nests can be traced back to the early times of the Ming dynasty (1368–1644). The story says that the Emperor sent a eunuch, Cheng Ho, with the imperial fleet on a series of missions to various countries in Southeast Asia. One of his main duties was to gather information about the type of food used by the local population in the visited lands. It is believed that Cheng Ho brought the first samples of edible nests to the court, and since then it has been a famous delicacy in China.

Demand for this prized commodity used to be so great that in the late 1700s some four million nests were traded from the Javanese port of Batavia, now Jakarta.

Demand for these nests has not diminished much even now, since the Chinese living in the USA alone consume about 30 tonnes of birds' nests annually. This means an incredible number of nests have to be harvested, as one nest weighs no more than 10 g.

The birds' nests are used exclusively for making a soup which is one of the most expensive and extravagant of meals. In Hong Kong restaurants one bowl of this soup may cost as much as $50.

Those who have the pleasure of consuming this meal seldom realise that the people who collect the nests risk their lives. In fact, gathering the nests is probably one of the most dangerous jobs in the world. In many regions, nest collecting is a traditional family occupation and men are trained for this job by their fathers from childhood. The nests are usually located more than 45 m above the floor in inaccessible caves on very steep cliffs. They are therefore very difficult to reach and require almost acrobatic skills.

To reach the high elevation a complicated scaffold is made from long bamboo canes. The gatherers' job is extremely hard and requires enormous strength and endurance, as they start climbing at sunrise and descend in the evening. Surprisingly, they do not eat anything while working, except the nests which they believe alone provide them with the necessary strength.

In some regions the gatherers believe that the caves are governed by spirits and they must therefore please the god who has power over their lives. So, before going to the cave they take offerings to a special shrine. They believe that bad spirits are everywhere, so even in conversation the collectors never use such words as 'fear', 'death' or 'blood'. They believe that the bad cave spirit might overhear, be offended and take revenge by bringing death during the nest gathering.

These gatherers obey certain other strict rules during nest collecting. For example they never use bare hands to collect the nests, but a particular tool which they believe has a special power to chase away bad spirits. It is hard to imagine a more risky job, since most forays are so dangerous that some collectors have said afterwards that they felt born for the second time. This is not an exaggeration, since some gatherers have died after falling from great heights while gathering nests.

People usually need to obtain special licences to gather nests and the job is usually performed by professionals. But the precious nests also 'invite' poachers. To protect them some caves are specially guarded. The entrances to the caves are even occasionally mined against potential poachers and warning signs given. There have been cases of poachers not taking the warnings seriously and being killed by stepping on the deadly mines.

While the most precious nests are those composed only of bird's saliva, the nests of another bird called the black-nest swiftlet (*Aerodramus maximus*) have also been used. These birds' nests are regarded as 'dirty' because as well as hardened saliva they contain the birds' feathers. These nests therefore fetch much lower prices.

These small birds also use their precious saliva for other purposes. They use it to glue together the many insects they collect to feed their offspring. These birds seem to realise that by sticking many tiny insects together the resulting ball is easier to carry to their nests. One such ball may contain as many as 500 insects. It is a big meal if one realises that the bird itself does not weigh more than 10 g.

In order to protect the bird population, the edible nests are only allowed to be gathered three times in a season. After the nests are gathered, the birds soon rebuild them and they usually use the same spots in the cave. The gatherers only collect the nests twice and the third nest constructed by the birds is always left untouched so that birds have a chance to raise their offspring. The gatherers only come back when the young birds leave their nests.

Although the raw nests seem to be almost tasteless, the Chinese blend it with chicken to make an excellent soup. They claim this meal is not only nutritious but also has a remarkable power to aid rapid recovery after illness and it became a wonder rejuvenating remedy for the elderly and an aphrodisiac. Until recently, however, the famous soup was regarded by most scientists as just ordinary food with no medicinal value. But recent research has revealed that the Chinese were right as there are substances in the nest material which may improve the power of the human body to fight disease as they seem to beneficially affect the human immune system. Some scientists are highly optimistic and even claim that these substances may be useful in combating AIDS. Let us hope that an increased interest in these little creatures will not lead to their total extinction.

These tiny birds live like bats in total darkness and since the caves may be several hundred metres long, the birds, it appears, should have a problem finding their nests. But they have no problem at all and find their way by using an echo location system. Their navigation skill, however, is not as subtle as that of bats. They are unable to catch insects in the dark and they occasionally lose their direction, bumping into nest gatherers' faces and frightening them.

As one would expect, the many thousands of birds continuously living in these enormous caves leave a colossal amount of faeces on the cave floor and all the caves are full of huge guano deposits. According to one explorer, the cave in which the tiny swiftlets live can be found with the nose much sooner than with the eyes, so powerful is the stench.

An unaccustomed person entering such a cave may faint. It is surprising that the famous delicacy enjoyed by so many people is obtained from such stinking locations.

7.
Mysterious Fathers

Unusual Deliveries

Frogs are among the most interesting of amphibians and the breeding habits of some of the species are truly unbelievable. An Australian native frog (*Rheobatrachus silus*), which was discovered in 1973, does not deposit its eggs, but swallows and incubates them in its stomach. Simultaneously the stomach juices, which would normally digest the eggs, are miraculously switched off. It has been revealed that the jelly surrounding the eggs secretes a special substance which inhibits the secretion of stomach acids. With such an incredible adaptation the eggs can be incubated safely for several weeks. Similarly, when the tadpoles hatch, they secrete this substance to prevent them being digested. Because of this strange habit, the frog mother does not eat during this entire period.

As the frog's babies grow inside this unusual incubating chamber, her stomach extends so much that it becomes as thin as a plastic bag. Eventually, the peculiar delivery takes place. The mother opens her jaws wide and vomits up her young straight into the water.

An unusual breeding habit of another frog, *Rhinoderma darwinii*, was discovered by Darwin during his famous trip on the ship *Beagle* to southern Argentina. As Darwin noticed with astonishment, it is the male frog which becomes the caring mother.

He approaches his partner when the eggs hatch and picks up the tadpoles and inserts them all into his vocal sac. To accommodate the tadpoles as they increase in size, the vocal sac expands enormously. When the young reach a proper stage of development, the father delivers them by spitting one after another into the water.

One would normally expect tadpoles to be smaller than the adults. But in one species, the paradoxical frog, (*Pseudis paradoxus*), the reverse is true. In this frog species, which occurs in Trinidad and in the Amazon region, the tadpole is a real giant when compared with the small adult frog. While the tadpole may be some 25 cm long, the adult frog is not more than

6 cm in length. It was initally thought that from such a huge tadpole an extremely large frog must develop and for this reason the tadpoles were considered to be a different species of amphibian. It is hard to believe that such a small adult frog could metamorphose from a more than four times larger tadpole. The common name 'paradoxical frog' is therefore fully justified for this frog species.

Long-Suffering Father

Penguins are perhaps the most human in appearance and most comical of all birds. They have an odd, upright stance and plump body and, in some species, a white 'shirt' front and immaculate black back, altogether resembling a gentleman in formal dress.

The largest penguins are the emperor penguins (*Aptenodytes forsteri*) which can stand nearly 1.2 m and can weigh some 35 kg. This bird is most famous for its ability to survive and breed in the antarctic winter when the temperature can drop as low as –75°C during very cold spells. It is hard to imagine that these birds can raise their offspring in freezing snow blizzards with penetrating winds that blow up to 200 kmh. No wonder that Edward Wilson, during his south polar expedition in 1907, could not believe his eyes when he saw an emperor penguin with an egg resting on its feet.

Surprisingly, what he saw was not a female holding her egg but a male, as it is he who takes the full responsibility for incubating the egg. Once an egg is laid, the male transfers it onto his feet and, by covering it with folds of abdominal skin, keeps the egg warm. This is in fact the only penguin species in which the male is responsible for egg incubation. There can be many hundreds of male birds huddling together to keep warm, most holding an egg on their feet. They form 'compact' circles on the ice called turtles. While balancing his precious egg, the bird moves around to find the best position to ensure that the egg is not exposed to the freezing air. Although the male must spend some two months like this he is constantly alert, ensuring that his egg does not fall off because otherwise another male without an egg may steal it from him. The urge to become a father is so strong among these penguins that occasionally fierce fights break out over egg possession in the rookery.

During the entire incubation period the father is so busy that he has no chance to feed and therefore when the female finally returns he is almost starving to death. She usually returns after the youngster hatches and she always brings plenty of fish for her partner as if in gratitude. If, however, the female partner does not come back and the young hatches the father feeds his baby by regurgitating a milky secretion from his stomach. When the female returns both parents care for a hatching baby and they huddle together to keep it warm.

Although the emperor penguin, like other penguins, is unable to fly, they are fantastic swimmers. Using their flipper-like wings, penguins can 'fly' in the water with a remarkable speed of up to 64 kmh. When walking on the ice, however, the emperor penguin usually waddles along the surface at only some 3–4 km an hour. In an emergency, however, the penguin applies another technique known as tobogganning. Lying on its stomach and propelling itself with its feet and wings a penguin can achieve a very high speed on the smooth ice surface.

Since the emperor penguin lives and breeds on floating sea ice it is the only bird which never touches land. As there is no building material on the ice, they cannot construct nests to facilitate incubation.

One of the great advantages of living in the cold antarctic regions for the emperor penguin is that these animals have escaped persecution by men who otherwise would have hunted them. In fact, many species of penguin, which inhabit warmer regions of the world, have been decimated by humans.

Some penguins contain large amounts of oil in their blubber to keep them warm, and they became an important source of oil, especially for lighting. The demand for this oil was so big in the past that in 1867 alone, one company produced 230,000 litres of penguin oil for which up to 400,000 penguins had to be slaughtered.

In addition the penguin's eggs attracted people's attention, as a food. In some regions where penguins were abundant, people could collect some 300,000 eggs in just one year in one area. No wonder that egg collecting became a popular activity in some regions. For example, on the Falkland Islands, schools were officially closed during the egg harvest. Children used to climb dangerous rocks to gather penguin eggs, often at great heights. More and more eggs were gathered every year and certain penguin species have now completely disappeared from around the capital city of the Falkland Islands.

It seems rather incredible but there are reliable reports that some penguins have their own graveyards. According to Robert Murphy, an American ornithologist, he saw a number of gentoo penguins (*Pygoscelis papua*) gathered around a pool, looking into the water. When the scientist returned to the same pool later, he saw with astonishment layer upon layer of dead penguins at the bottom of the pool. It seemed as if the penguins had committed suicide. Despite this observation, there is still no scientific proof that penguins commit suicide when they become old or sick by jumping into a pool.

Pregnant Father

The seahorse is one of the most peculiar fishes in the world. It has the head of a horse, the tail of a monkey, the outer skeleton of an insect and the independently moving eyes of a chameleon. For many people, it looks like a knight in the game of chess. Even the way it swims is peculiar as the seahorse swims in an upright position, swaying from side to side. Ancient Greeks must have been impressed by this unusual fish as they called it 'hippocampus' meaning 'the sea-monster horse'.

Seahorses have always fascinated people. Poseidon, the Roman god of the sea, was always portrayed riding in a chariot pulled by a pair of seahorses. In an old Turkish tale, a young princess used her faithful seahorses to conquer a faraway mythical kingdom.

Because of their strange appearance, seahorses were believed to possess a miraculous power to cure many ailments, and Pliny recommended the ashes of seahorses as a cure for baldness, and for the bite of a mad dog. The belief in seahorses' special power has survived till present times and dried seahorses can still be purchased from pharmacies in China. The creature is used both as a love charm and as an aphrodisiac. In some regions, dried seahorses are made into ornaments, often covered with silver or gold, and are believed to bring good luck, especially for attracting lovers. For this purpose powdered seahorses are mixed with drinking water and given to real horses in Java as a tonic.

Probably the most intriguing thing about seahorses is their breeding habit which has always interested naturalists. It is not the female but the male which gives birth to the young. The male approaches the female and tries to attract her with his dance performance. When his invitation is accepted they come close together and touch bellies. During this loving embrace, the female inserts her cloacal appendage into the male's brood pouch and squirts all her eggs into it. In this unusual way the male becomes 'pregnant' while the mother swims away completely relieved of any parental duties. If the pouch is not full, the male invites other females to deposit their eggs. As a result the male's pouch swells up like a balloon, but the father nevertheless appears to be unperturbed. He incubates the eggs for about 45 days.

When the eggs are about to hatch, the pregnant father wraps his tail around some seaweed for support. He then begins to give birth to his babies. By contracting his pouch muscles he ejects his tiny youngsters, usually in ones or twos. This is a most difficult and drawn out delivery, lasting up to two days as the number of babies can be astonishing. According to Amanda Vincent, who studied the behaviour of seahorses, the 'mister mother' as she called it, can deliver as many as 14,000 tiny babies.

The tiny youngsters which are only a few millimetres long are almost exact replicas of their parents. Their bodies are transparent so their tiny beating hearts can easily be seen. Being transparent in the early stage of life makes them almost invisible in the sea water and hence less vulnerable to attacks by predators. Despite this, not many reach adulthood.

The delivering father is sometimes so hungry that he swallows some of his numerous tiny babies if they happen to come near his mouth. Unlike kangaroos, the seahorse does not allow his babies to use his pouch again even in an emergency. The male seahorse's pouch is just a temporary arrangement, designed only for incubation and delivery.

Seahorses are very poor swimmers, in fact they are the slowest fishes in the world, and for this reason they are unlikely to save their lives by swimming away from predators. To remedy this, the seahorse has developed an excellent camouflage. It can change colour and by doing so remains almost unnoticed amoung the seaweeds where it lives. The seahorse can adopt a great variety of colours to protect itself. The common seahorse (*Hippocampus hudsonensis*), which occurs in the Atlantic Ocean, is known to produce spectacular colours at will, changing from white to green or to bright red.

Some seahorses do not rely solely on changing their colouration to match their surroundings for camouflage. An Australian seahorse, *Phycodurus eques*, matches the colour and appearance of a seaweed with beautiful green colouration and in addition surrounding its body with leaf-like outgrowths. This remarkable seahorse, which resembles a piece of floating seaweed, is called the sea dragon.

Some seahorses, found in Australian seas, may reach an enormous 30 cm in length and are real giants compared with the other hundred or so much smaller seahorses found throughout the world.

Although most fish are silent, seahorses are known to produce a distinct 'clicking' voice, especially during courtship. Some biologists suspected long ago that seahorses use voice to communicate with each other. In 1939 a Japanese scientist reported that two seahorses, kept in separate aquariums, 'conversed' with each other as if using a telegraphic code. Recent studies have shown that when seahorses are kept in separate aquariums, they make a characteristic call to each other at regular intervals. Perhaps further studies will reveal that these odd sea fish have even more fascinating habits.

The Crucifix Fish

It is hard to believe that certain fish's bones could be more valued than their flesh. For the marine catfish of the *Arius* found in South America this is so. These fish's skulls have a peculiar structure. When cleaned, the skull shows an unusual resemblance to the figure of Christ with outstretched arms. Both sides of the skull are most peculiar. The most interesting is the lower part where the paraspheroid bone forms a figure strikingly similar to the body of Christ. There are two projections which can be seen as Christ's outstretched arms. The Weberian bone forms a 'halo' above the figure's head. But this is not the end of the story since the upper side of the skull can be seen as a Roman soldier and the dorsal spine as the sharp lance point with which Christ's side was speared.

Small wonder that these extraordinary skulls, which represented the Saviour's Passion story, became objects of veneration for devoted Christians in South America. Painted and mounted, it made an attractive icon and this famous fish became known as the crucifix fish in South America.

Some catfish have become famous for their fascinating breeding habits. In sea catfish of the family Ariidae, it is not the female but the male who becomes the real mother. Just after laying her eggs, the female transfers them to the male's mouth where they are incubated. Since the eggs are rather large — some 2 cm in diameter — the male's mouth must stretch enormously to maintain the thirty or so eggs. This living incubator must become even larger when the youngsters hatch, since they are about 5 cm long and so need more space. Fortunately, delivery soon follows, and the almost 'bursting' male starts spitting his babies forcefully into the water. During the whole incubating period, which takes about four weeks, the male does not eat.

There are many more peculiarities among the some 2,400 species of catfish. There are some, for example, whose large swim-bladders vibrate producing a distinct sound. A catfish *Doras maculatus*, which occurs in South America, produces such a loud sound inside its body that when it is taken out of water it can be heard from a distance of over 30 m.

There are catfish which emit a variety of voices. Some squeal and growl and there are even those that miaow like a domestic cat. Most catfish have the characteristic barbels known as 'whiskers', hence the name catfish.

These fish vary greatly in appearance and size. The largest is the European or wels catfish (*Silurus glanis*) which occurs in fresh waters in Europe and is a giant among catfish. One fish which was caught in the Dnieper River measured 4.5 m and weighed some 300 kg. Such a huge fish can even be dangerous to humans. According to a sixteenth century report, a child's head and parts of its hand were discovered in the stomach of a giant catfish.

Some catfish are so small that it is hard to believe that they could be a similar species. In fact, a tiny catfish called candiru (*Vandellia cirrhosa*) is usually no bigger than a match stick. This tiny fish, which lives in some rivers in Brazil, is not only a biological curiosity. It is known to attack humans. The tiny fish has a nasty habit of wriggling into the urethral openings of both females and males bathing in the rivers, especially when they urinate in the water. These fish have backward-pointing spines, so they cannot be removed without a surgical operation. They cause excruciating pain and are hated by the native people of Brazil. In order to protect their bodies, people entering the rivers where these fish are known to live, wear special guards made of palm fibres to cover their genitals.

Even more dangerous is the colourful marine catfish *Plotosus anguillaris*. This fish is up to 50 cm long and has conspicuous longitudinal black and bright yellow stripes as if to warn its predators. It is in fact one of the most poisonous fish of the Indo-Pacific Ocean. Its dorsal and pectoral fins are equipped with sharp spines, but unlike other stinging fish whose sting is inflicted with pressure, a slight scratch from this fish is enough to cause a severe sting. When touched, the spine springs out, injecting venom into the victim's flesh. The pain is so excruciating that it may cause severe shock and even collapse in sensitive people. A severe sting may be so dangerous in some people that death may follow.

Thermometer Bird

When the first European settlers penetrated the Australian scrub, they noticed some strange hillocks. Initially they thought that these structures were Aboriginal tumuli. When the natives told them that these mounds were made by birds, they could not believe it.

Similarly, American soldiers fighting in New Guinea were once fooled by peculiar structures which looked man-made. When they saw such a big mound, they concluded that it was the work of the retreating Japanese troops burying their military equipment. They were furious when, after a few hours digging inside the mound, they only found a few empty egg shells.

Unlike most birds which make 'ordinary' nests to incubate their eggs, these birds construct an extraordinary, huge mound. Among the thirteen species of these unusual birds, the best known is the Australian native mallee fowl (*Leipoa ocellata*) which is a large bird, the size of a turkey. This bird can construct a mound up to 1.5 m high and some 5 m in diameter. To build such a huge mound the birds must collect at least 50 tonnes of plant material and soil. The work is mainly the responsibility of the male while his 'spouse' only occasionally assists her partner. With his sturdy legs, he carries pieces of dead vegetation often from a radius of some 50 m. He seems to work according to a plan. He first digs a hole in the ground about 5 m in diameter and about 1 m deep and here he deposits a mass of dead leaves, twigs and pieces of bark.

When the rain comes the plant material begins to rot and the male then covers it with a layer of soil, making a mound up to 1.5 m high. The bird seems to know that the rotting vegetable matter will become a source of heat for his incubator. But the temperature inside the mound can be too high and so the clever bird does not allow his 'wife' to lay her eggs until he has carefully checked the temperature. The bird behaves like a trained laboratory assistant as he measures the temperature inside the mound. He inserts his open bill into the mound so that his highly sensitive tongue can register the temperature. Incredibly, the bird is known to detect changes in temperature with remarkable accuracy in the range of a tenth of a degree. After 'reading' the temperature, the bird reaches a conclusion and reacts accordingly. If he finds the temperature is too high he immediately starts removing some of the soil covering the mound to dissipate excess heat. If the temperature is too low he covers the mound with additional soil. Once he finds that the conditions are perfect, he makes a tunnel and invites his female partner to lay eggs in the incubating chamber inside the mound. After laying her eggs the female leaves the chamber through the same tunnel, and her 'husband' closes it carefully. From then on, the male checks the temperature inside the mound daily and is able to keep it at about 33.5°C, the ideal conditions for the incubation of eggs. Biologists were always puzzled by the male's remarkable ability to keep a steady temperature inside the mound. When an Australian scientist tested this by placing an electric heater inside the mound, he was surprised by the bird's immediate response. It started to remove the soil from the top of the mound so that the excess heat could escape. Soon the temperature dropped down substantially.

Since the female lays her 16–33 eggs singly at intervals of up to a week, the total incubation period is extremely long, and the male is busy for several months working hard to maintain the proper conditions within the chamber. The male mallee fowl seems to be able to foresee changes in the weather. When a storm is imminent the male quickly reacts, covering the mound with additional soil as if realising that the storm may destroy his whole construction.

Once incubation is over, the hatched chicks must work for hours to dig to the top of the mound as the egg chamber lies some 1.5 m inside the mound. The hard-working parents take no more interest in their chicks, so the youngsters receive no help after birth. Fortunately a few hours after hatching, the young are already able to run and escape into the scrub.

First European settlers were unfamiliar with this bird's habits. An amusing story is told of a farmer's wife who was once given a few mallee fowl eggs by friendly Aborigines. She placed them in a hat with the aim of rearing her own chicks. On her long way home, she was surprised to see one of the eggs crack open and a chick drop to the ground from the carriage and run off into the bush.

The Australian Aborigines find these birds' eggs delicious and collect them, spending many hours digging up the huge mounds. Their efforts are sometimes successful since about 30 large eggs can often be found in a single mound. Each egg may weigh some 180 g so they are valuable food.

Even foxes living in the bush have learned how to dig up the mounds and collect the eggs. Egg collecting is one of the reasons for the considerable decrease in population of these remarkable birds in Australia. But even more important is the clearing of the forest which destroys the birds' habitat. Fortunately, in recent years, many central and southern Australian farmers started to realise that further clearing of the bush would lead to the extinction of the unique bird of which Australians are proud. Some farmers have already shown their respect for this bird by deliberately leaving certain parts of their mallee scrub uncleared in order not to disturb the mallee fowls living there.

8.
Keys to Paradise or Hell?

Narcotic of the Mayas

In many cultures, 'ugly' toads were associated with horror and their appearance signified a forthcoming tragic event. It used to be believed in parts of Europe that if a toad was molested it would come back at night and spit saliva into the offender's eyes, causing a deadly disease. There is some truth in such a belief since many toads are exceedingly poisonous.

When the early Maya cities were excavated in Mexico the archaeologists were astonished to discover heaps of toad remains near the temples. They were marine toads (*Bufo marinus*) which are common in Central and South America. It is assumed that the ancient Mayas must have discovered that these highly poisonous toads could be used to alter the human mind. They must have learned how to extract their poison and use it not to kill, but to induce a state of wellbeing. Such a narcotic was an important addition to their religious rituals when a human victim was sacrificed at the temple altars.

Under the influence of the narcotic toad venom, the chosen victim obeyed the ceremonial priests' commands as if immune to the torture of the cruel sacrifice. It is believed that those who took part in the often macabre rituals also used the hallucinogenic extract.

The marine toad is very interesting. It is in fact the largest toad in the world since it can be up to 24 cm long and weigh over 2 kg. It is a typical toad shape, with short, broad body, bumpy skin and hooded eyes. Its poison glands, which appear as warty lumps are at the back of the head and elsewhere. If a toad is molested the poison oozes out onto the surface of the skin. It can squirt its poison in a spray for a distance of up to one metre. If this venom is swallowed, often by licking contaminated fingers, serious poisoning may occur.

The giant marine toad has an interesting history, not only in ancient Mexico, but also, more surprisingly in contemporary Australia, where it became known as the cane toad. The toad was 'invited' to this country as it was believed that it could help farmers control the population of certain beetles which decimate the sugar cane crops. Marine toads had already been 'employed' successfully for this purpose in other regions of the world, so the idea seemed reasonable and found general support.

A hundred marine toads were imported to Australia from Hawaii in 1935 and produced some 1.5 million eggs. About 60,000 hatchlings reached adulthood and were soon released into selected sugar-cane plantations in Queensland. Although there was much enthusiasm among farmers and expectations were high, the whole programme was actually a miscalculation. The cane toads did everything except eat beetles. They found the cover insufficient and tried to escape from the heat of the tropical sun. Having no natural predators, they started to multiply at an enormous speed. No wonder, because the male's sexual urge is so strong that a hundred males may fight between themselves over one female. A male was even observed mating a female that had been dead for some time.

A female toad can lay some 35,000 eggs in just one year, thus, not surprisingly after a few years, toad numbers increased enormously. This toad can also eat almost anything and has no predators. After a few years the marine toads became a real pest and the situation in some regions became so alarming that immediate action was urgent. In some regions of northern Queensland $30 was paid for a single marine toad captured or killed. During so called 'toad eradication nights' thousands of toads were killed by volunteers who walked for many hours and used long sticks to beat the 'horrible' toads.

The toad hunters were given special pamphlets which informed those interested how to make their job more 'merciful': 'You take a toad, stick it in a plastic bag, do up the plastic bag and stick it in the freezer, and the toad just goes to sleep'. The authors of this pamplet did not realise, however, that the toads actually suffered a lot because in the fridge it would take them up to 30 minutes to die.

Despite the eradication programmes, however, the battle with marine toads is far from being won. They still occur in such masses that often driving a car without squashing them is almost impossible. Passing vehicles are believed to produce some 200 tonnes of toad flesh on the roads in Queensland a year. The dead and rotten marine toads make such a stench that it has been compared to a 'swarm of mullets coming from the river'.

Marine toads have also created another problem in Australia. Some people seem to have 'rediscovered' the hallucinogenic properties of their venom, as they started to use it as a narcotic. Like the ancient Mayas people wanting to alter their minds have become addicts of this new and easily available narcotic. A number of people, especially the young, started to produce their own hard drug using the marine toad toxin, killing them only for this purpose. The Drug Squad in Queensland found that there were certain preparations on the black market, even labelled 'hallucinogenic cane toad skins', which were claimed to have lasting hallucinogenic effects when put into a pipe and smoked.

The venom contains a very potent substance called bufotenine which can cause an addiction similar to that of other hard drugs such as heroin. In order to discourage people from making home toad 'extracts' the cane toad preparations were made illegal in Australia. Toad venom is now listed as a dangerous hallucinogenic drug in Australia and a severe penalty for its possession or distribution has been introduced.

Strangely, the 'ugly' marine toads have found devoted friends among some Australians who even keep them as pets. Gordonvale in Queensland became famous because some children in this town kept them as dolls. Dressed up in clothes, these rather strange dolls had their own comfortable beds to sleep in and were taken for walks by the devoted girls. Keeping such dolls should not be encouraged because if babies or small children lick the toad's skin they may become seriously poisoned and may even die.

Russian Roulette

To deter predators certain marine fish, called pufferfish, employ a peculiar technique. When under imminent attack, the fish rapidly inflates its body into a grotesque balloon which may be three times larger than its original size. Some of these fish have spines which are normally flattened, but when they inflate, the spines rise and the fish look even more frightening. Predators are usually frightened by the sudden change and do not attack.

The pufferfish has a special adaptation for performing such a trick at will. The muscles of its gills and stomach, as if on a command, begin pumping water incredibly fast so that its body swells. When the threat is over, the fish expels the water through its gills, mouth and even anus and returns to its former shape. Even if taken out of water the pufferfish uses the same trick, but uses air this time to inflate itself. It is able to reinflate itself several times in succession.

Pufferfish possess another powerful weapon against predators — an internal and potent poison which kills the creatures which eat them. They seem to warn predators by being vividly patterned and coloured.

Pufferfish, which are found in many seas, especially the Indian Ocean and the South Pacific, have long been known for their deadly flesh and natives along the Hawaiian coasts used the fish gall to poison their spears.

Despite the fact that many pufferfish contain poison which is over 20 times as powerful as the famous curare and many people have died after eating such fish, the Japanese have an entirely different attitude. They regard it as so delicious that they do not hesitate risking their lives to eat it. In the past whole families in Japan have been killed as a result of eating this fish they call 'fugu'. A strange cure used to be applied against fugu poisoning. A poisoned person was buried up to the neck in earth in the belief that this would eliminate the poison. The cure was usually ineffective. To eliminate the poisonous substances, specially trained chefs spend much time trying to make their fugu safe. But before such a technique was known, many people became victims of their gastronomic passion. Some victims are said to have recovered while awaiting burial or before cremation as they wake up from the fugu's paralysing effect. They said they were conscious and could even recall many comments made by the doctors and their close relatives after they were pronounced dead.

A safe fugu soup is among the most expensive dishes in the world as it may cost as much as $150 for a single serving. It is said that many eminent politicians and personalities in Japan are 'addicts' of fugu. The famous Japanese actor Mitsugoro Bando VIII, who was a great lover of the hazardous fugu, died after eating some fugu liver in one of the less respectable restaurants. Some people think fugu eating is a kind of Russian roulette, but most Japanese consider this meal most exciting and worth trying. While some of them say that fugu is like their paintings, that is delicate, subtle and elusive, others have another saying expressing their philosophy towards fugu: 'Those who eat fugu are stupid. But those who don't eat fugu are also stupid.'

It is very difficult to make fugu safe to eat, because the poisonous substance occurs in all the internal organs such as the intestines, liver, sexual organs and even the blood. This is why a person who prepares fugu must undergo a long training, after which he works like the best surgeon.

The deadly property of certain pufferfish was well known to shamans or medicine men in Haiti who used the fish's flesh to prepare the most incredible poison ever discovered. Mixing this poison with certain plant extracts such as *Datura* species, they made a special poison which changes a person into a zombie, or 'living corpse'. Stories of zombie making were formerly reported by anthropologists who were told that some people, who died as a result of sudden illness, reappeared in other regions although their graves seemed to remain intact. A true case of zombiefication was reported in 1983 by a scientist working in Haiti. The scientific world has learned with astonishment that Haitian shamans use a poison to make a man seem dead. The man, who is in fact in a deep coma, is pronounced dead and buried, but his body is secretely removed from the grave and resuscitated, but being confused he agrees to be sold as a slave. Such zombies do not return home in a belief that their sudden appearance may frighten their relatives to death.

The fantastic poison, when used 'correctly', transforms a healthy man into a zombie, which seemed incredible in the past. The pufferfish used in the preparation of the zombie poison are of various *Diodon* species. They all contain a powerful poison called tetrodoxin. The Japanese have learned how to isolate this poison in a pure crystal form from the ovaries of certain pufferfish. Highly diluted, this substance has been used as an effective pain killer especially for terminally ill cancer patients. Tetrodoxin is one of the most expensive natural products as one gram costs about $50,000. It is strange that a poison which has been responsible for many people being buried alive, has become in Japan one of the most beneficial pain relievers known to medicine.

Small wonder that this most unusual fish has become so famous in Japan that a statue of a fugu fish has been erected in a park in Tokyo. A famous temple near Osaka has a memorial tombstone honouring fugu.

9.
Great Athletes

The Champion Frog

Frogs are among the finest distance jumpers and for this ability they have been admired since time immemorial. With their well-developed hind legs frogs are born for jumping and some are able to make tremendous leaps which are really impressive. The best jumping frog in the world is the 5 cm long sharp-nosed frog (*Ptychadena oxyrhynchus*) which occurs in South Africa. One of these frogs holds the record for the greatest distance jumped as it covered 5.3 m in a single leap. This means that it jumped 100 times its own length, which is equivalent to a man jumping some 170 m.

There are many stories about the proud owners of frogs which were unbeaten, 'terrific jumpers'. Some people who organised contests made a lot of money with their remarkable pets. The most famous is the story described by Mark Twain. According to this story, there was once a man in California who owned a champion frog which always won contests with other frog 'athletes'. One day a stranger came and said he did not believe in this frog's remarkable jumping ability and if another contestant was found he would prove it. The owner went away to find a new frog, leaving his champion with the stranger. When he returned, the contest was held and to his surprise the new frog jumped well, while his 'champion' did not even move. He lost the contest and paid the wager which the stranger took and ran away. When the owner inspected his frog, however, he found a number of heavy lead pellets attached to its belly.

This story became so popular that to commemorate it a special Frog Jumping Jubilee was organised in Angel's Camp in California. The contest was so popular that it became an annual event with many participants. Each frog is allowed three jumps and the frog which covers the greatest distance in three jumps wins. Special rules are observed to eliminate cheating. Since frog jumping competitions are held in many parts of the world there is even an 'International Board of Frog Jumping' resident at Angel's Camp which is the main regulating body of this peculiar sporting event. Some people are so devoted to betting on this sport that it is said that fortunes are sometimes lost, but rarely won.

Frog jumping competitions have also been known to take place in Australia and in 1973 there was even a controversy about rules. It has been established that some dishonest competitors doped their frog athletes by forcing them to drink whisky to make their performance more spectacular.

One of the poorest jumpers among frogs is the goliath frog (*Conrana goliath*) as it can only cover some 3 m in a single leap. This huge frog, which can weigh over 3.5 kg and attain the length of some 36 cm, is too heavy to be a real athlete. It can only make three to four consecutive leaps and must then stop because it is so exhausted by this effort. This is a very impressive looking frog and it only occurs in the waterfalls of Cameroon and Equatorial Guinea. The African natives likened this frog to a child because of its legs. They call it 'niamoa' which means 'mother's son'. It is rather surprising that such a gigantic frog is not greatly relished by them. Only the Pigmies are known to hunt these frogs in the jungle, while other local inhabitants seem reluctant to harm them.

Fantastic Sportsmen

Fleas hold the world record for high jumping. The best jumper is the common flea (*Pulex irritans*) which leaps to a height of about 20 cm. Proportionally, a man would have to jump over 200 m, the height of an eighty-storey building, and land on his legs unharmed. The flea's leg and thorax muscles are specially equipped with a superelastic protein which enables the insect to catapult with such enormous power. Its jump acceleration is believed to be 50 times faster than that of a space shuttle's launch. Fleas seem to love jumping. A specimen of one species was observed to make 30,000 consecutive jumps without resting.

Fleas are also remarkably strong and some can tow objects over 200 times their own weight. The incredible ability of some fleas has not been overlooked and people in the past used fleas for entertainment purposes. The tiny insects became performers in the world's smallest circuses. Only human fleas were suitable and some circus owners used to import them from Russia, considering them to be the best.

It was not an easy job to turn a flea into an artist because of its habit of continuous jumping. To handle the tiny creature, a minute silk thread collar had to be attached to its body. Minute chariots with real wheels were specially made and propelled by the fleas. A chariot with a miniature coachman in full dress, drawn by tiny flea horses made a great impression on the audience. They were even 'taught' to play in a tiny orchestra. Their legs were hitched to minute chairs with instruments attached and when the fleas started to wave their legs, the instruments produced tunes from the attached musical box. Although the owners of the famous flea circuses claimed they trained their tiny actors to perform their 'incredible' tricks, their movements were in fact the flea's efforts to escape when restrained. Even special chariot races with fleas harnessed to the vehicles were organised in the past in Europe and there was always a demand for such entertainment. The most famous was the Munich flea circus where a great variety of flea performances could be observed, including flea dances. Some of these flea circuses survived till recent times. The small artists had to be fed with human blood and this was a small sacrifice made by the circus owners themselves. They used to attach their fleas to their skin in a special tiny box so that the fleas could suck their blood.

It is hard to believe, but the tiny flea has been responsible for the death of many people. It caused the outbreak of the bubonic plague which killed some 25 million people in Europe between 1347 and 1352. As the cause of this dreadful disease was unknown, people suspected that poisoned water was to blame. People even started to suspect that such massive poisoning was caused by the Jews and so thousands of innocent Jews were dragged from their homes and killed. But the epidemic did not cease. For the Church authorities, the plague was God's punishment for humans' sins. A committee of doctors of the University of Paris claimed, on the other hand, that the disease was the result of the conjugation of Saturn, Jupiter and Mars.

Despite many efforts to find the cause of this disease its origin remained unknown for centuries. It was Dr Paul-Louis Simond, a French scientist, who was the first to notice that the rats he dissected carried the dangerous bacteria causing the bubonic plague. On 2 June 1898 he declared: 'I have just unveiled a secret which has tormented man for so long.' We know now that although bacteria causing the disease live in the bodies of infected rats, it is a flea which transmits the disease when it bites people after biting such a rat. When a flea bites a man it transfers the bacteria and this was, in the past, a deadly bite.

With the use of insecticides dangerous flea infestations can be controlled more efficiently and, as a result, epidemics of bubonic plague are now very rare. Even people infected with this disease now have a good chance of surviving if antibiotics are used.

Not all fleas are dangerous to man. Out of over 2000 species and subspecies of flea only some 120 are believed to be able to transmit the disease, and among those only some 20 species bite man. But even a 'normal' flea bite was and still remains a problem in filthy, unsanitary areas.

The ancient Egyptians invented a rather peculiar method for getting rid of fleas. Wealthy people used slaves for this purpose. They covered a slave with a mixture of asses milk and blood and all the fleas in the house were said to gather on his skin. In seventeenth century Europe special flea traps were commonly used, especially by women who were said to be more vulnerable to flea bites. These were hollow cylinders made of wood or ivory into which a cloth smeared with blood was inserted to attract fleas. Sometimes, instead of blood, a rod was smeared with sticky honey and when the fleas entered the trap they became stuck. Such traps were usually hung on women's breasts. Flea traps were so commonly used that producers even made traps with a small microscope attached to the set so that the captured insect could be easily observed.

Fleas are found nearly everywhere, and they can even live in polar regions. The fleas can be frozen in the ice for several months and revive. Depending upon the species they can live from several weeks to up to four years. Some fleas have such an amazing capacity to survive that they can live without food for months.

In many flea species, when a young flea hatches from the cocoon, it can copulate almost immediately, and the female can start laying eggs just a few days after birth.

Allah's Gift

The Arabs of the deserts call the one-humped or dromedary camel (*Camelus dromedarius*) 'Allah's greatest gift to mankind'. So incredible is their devotion to this animal that there are over 5,000 different names and epithets for the camel in Arab literature. Many famous Arab tales refer to the camel. According to one, when Mohammed received the revelation of Islam, his camel fell down on her knees under its weight. And when Mohammed revealed the hundredth name of Allah to his camel, this secret message was immediately passed from one camel to another, which is why all camels have such a proud and disdainful expression. The Arabs also explain the origin of the camel's peculiar 'smile' by saying that when the male camel learned that God was going to create a spouse for him, he smiled so broadly that his nose split and has remained so ever since.

This marvellous beast seems to have been created to satisfy all the needs of an Arab living in the desert. Camels are not only excellent beasts of burden, but they provide the Beduins with nutritious milk and meat. The camel's leather is made into sandals and containers and its hair is made into rugs and cloaks and the cloth from which the Beduin's tents are made. Camel's droppings are an important source of fuel in the desert.

Amazingly, even the contents of the camel's stomach can be of importance. When a Beduin is thirsty and no water is available he has his camel to drink from. He forces the animal to vomit and the liquid obtained can be sufficient to prevent complete dehydration and save a man's life in an emergency.

The Beduins have many other interesting uses for camel 'products'. Camels' vomit mixed with a little urine has been used to alleviate stomach problems. The camel's urine is used as a cure for skin diseases and for cleansing wounds. In the old days, camel urine was used as a hair shampoo, and it was even occasionally used in desert regions to wash new born babies. The camel's milk mixed with honey is regarded by Beduins as a sexual stimulant.

The camel is so precious among the Beduins that it is considered an important gift and it is even used instead of money in various transactions. Camels can be offered as reparation for an injury or even murder.

Camels are excellent animals for carrying heavy burdens and, in contrast to other animals, they will keep marching until they drop dead. Trade, in the old days, without the use of camels is difficult to imagine. Caravans of thousands of camels loaded with precious myrrh and frankincense used to wander along the desolate Incense Road in the Arab Peninsula. Such caravans were often attacked by marauders called 'karvan', from which the word 'caravan' is derived. Even today thousands of camels can be seen carrying salt to remote and desolate regions in the southern Sahara. Sometimes more than 20,000 heavily laden camels carry salt from Taoudenni to Timbuktu. The whole trip takes three weeks and the salt is exchanged for millet which is then carried back.

Since ancient times, camels have been used in warfare, and the first report of camels in battle dates back to 190 BC.

When the Goths reached the Danube basin in the fourth century, as well as horses, they used a large number of camels brought from Mongolia. The use of camels in warfare became even more common after 1772 when the Afghans introduced a light pivot so that a soldier could fire a cannon from the animal's back. This strange cavalry was successfully used against the Persian army.

When Napoleon invaded Egypt in 1798 he also used camels instead of horses, and he himself rode a camel. He was so devoted to his camel that when the campaign was over, he took it back to Paris. Napoleon's camel became very famous and when it died it was stuffed and exhibited in the Natural History Museum in Paris for over one hundred years.

Camel cavalry became so important in battle and used so extensively that during the Second Afghan war (1878–80) the British troops lost some 70,000 of these animals.

Camels were also introduced to Australia where they were extremely useful for penetrating the vast desert regions of this continent. Many famous explorers used camels on their trips which would have been impossible to make without the 'help' of camels. The first camels were brought to Australia with their Afghan drivers in 1840.

Strangely, Australia is the only country in the world where there is a large feral population of dromedary camels. Some 10,000 of them are estimated to wander throughout the semi-desert regions of Australia. In some areas, wandering camels invade crops to such an extent that farmers are allowed to shoot them on sight to protect their fields.

It is not only the camel's remarkable history that is of interest. The animal itself is unique in its habits. They seem to be created especially for the desert conditions as they have a fantastic adaptation to extreme heat, dust and lack of water and food, conditions in which no other mammal could survive. The most remarkable is the camel's ability to survive for up to ten days without water. The animal may lose up to 30 per cent of its body weight, and when water is available it will gulp up to 130 litres in just 10 minutes to fill its 'tank'. It has this enormous capacity to live without water because it loses a very small amount of water as sweat and urine. In fact much of the camel's urine passes through the liver back into its stomach where it is processed into proteins and so only a very small amount of urine is excreted through the kidneys.

Although some people believe that the camel stores its water in its huge hump or humps, it is not true as the hump is a storage room for fat only. This fat is its main source of energy when normal food is not available during the long march through the desert.

A peculiar double row of heavy lashes efficiently protects the camel's eyes from flying sand and grit particles and his slit-like nostrils can be closed to prevent windblown sand penetrating his lungs. The large, fleshy pads on his toes prevent his feet from sinking into the desert sand.

The camel is extremely resilient and can carry a load of up to 200 kg and cover a distance of 100 km a day in the most harsh desert heat. Without food and water, the camel will still keep going.

Although often described as the most stubborn of animals used by men, camels are in fact men's most reliable partners. For an Arab travelling in the most desolate regions his camel may be his only company for several months. No wonder that some camel owners show a great devotion and even love to their beasts of burden. There is an amusing proverb among the desert Arabs that 'The best of women is a frolicsome she-camel'.

The second species of camel is the two-humped bactrian (*Camelus bactrianus*) found in Central Asia. It endures the most inhospitable climate of the Tibetan Plateau where in summer the temperature reaches +60°C and in winter drops down to arctic cold.

A camel may not be very attractive when observed close up, but when hundreds of these animals walk in a row it is a remarkable sight. The animals walk with a peculiar swinging stride, which results in a characteristic rolling effect and this is one of the reasons why the camel is called the 'ship of the desert'.

10.
Crooks Among Animals

Playing Possum

Opossums (family Didelphidae) were the first marsupials encountered by the western world. They were discovered in 1500 in South America by Pinzon, the Spanish explorer. When this animal was presented at the Spanish court, it caused a great sensation with its rat's face, the prehensile tail of a monkey and naked, bat-like ears. Its most amazing feature, however, was its large sack for carrying its baby.

The only North American opossum is the common or Virginian opossum (*Didelphis virginiana*) which also occurs in South and Central America.

Strange beliefs are associated with this animal in some regions of the USA. People used to believe that the female gave birth through her nostrils. This peculiar myth was probably based on an observation of a female opossum holding her baby in her mouth. These opossums are particularly famous for their peculiar method of defence. When an opossum is under attack with no chance of a quick escape, it uses a clever trick to fool its predator. It falls on the ground immediately, with its eyes shut and tongue hanging out. It becomes motionless, and its heart beat and breathing appear to have ceased completely and the animal seems to be dead. It shams death so perfectly that the predator loses interest in attacking a 'dead' animal. But once the danger is over, the opossum immediately comes back to life and runs off. The trick is so effective that even some hunters have been fooled. When they have carried an apparently 'dead' animal by its tail, they have sometimes been severely wounded by the suddenly 'resurrected' animal.

This peculiar self-defence technique has always puzzled biologists. Most of them, however, did not consider it to be a deliberate trick used by the 'clever' animal but rather sudden shock which caused a kind of 'coma'. Recent investigations have revealed, however, that during the animal's 'death' period its brain works normally, and that the animal is in a state of constant alert. This confirms that the opossum is really 'playing possum'.

The Virginian opossum has the shortest gestation period among mammals since it takes only eight to thirteen days, which is very unusual. The female may produce as many as twenty babies at a time but only thirteen of them will survive. This is because she has only 3 teats in her pouch. Only those babies which are quick enough to grasp a teat can feed themselves. Each baby fixes itself to a teat and remains fastened to it for several weeks, so it cannot share the teat with other babies who starve to death.

When they are about 10 weeks old, the youngsters become too large for the pouch and they have to leave. They climb on to the mother's back where they entwine themselves in her long fur until they are ready to start their independent life at the age of about three months.

Among the opossum species found in South America, there is one which has chosen an aquatic life. This is the water opossum (*Chironectes minimus*) which lives in the Amazon and other large rivers of South America. It is nocturnal and it spends the day in burrows near waterways. This animal is well adapted to life in the water, and the female can dive with her children inside her pouch.

She has special muscles which tightly close her pouch, so when swimming under water, the shelter is closed and her babies do not drown. The youngsters can breathe for some time using the air supply in the pouch.

The name 'opossum' has also been applied to the primitive marsupials found in Australia. These arboreal marsupials are now called possums to distinguish them from the American animals.

Many Australian possums are famous for their ability to 'fly'. The most interesting is the greater glider (*Petauroides volans*) which glides by spreading its ventral membranes to form a kind of a delta-wing aircraft. Its long and bushy tail is also an excellent aid in its navigation. It usually 'flies' from one tree canopy to another and if it starts its flight from a very tall tree, the animal may cover some 120 m, quite a long distance for a mammal in the air. Strangely, during such a flight the possum may carry a passenger. The glider's baby may sit on its mother's back by clinging tightly to her long fur.

'Honest' Thief

The packrat (*Neotoma* species) found in North America has an extraordinary habit of collecting various objects and hoarding them in its nest. This small rodent, also called the wood rat, steals shiny objects such as nails, pieces of glass, a small watch left on a table and some are even known to have pilfered false teeth from a glass. This astute thief would even take coins from the purse of a sleeping owner.

Early American settlers, who became acquainted with the peculiar habit of the packrat, called it a trade rat as they were convinced that this clever animal always trades and never steals, as it always leaves other objects in their stead.

There are many amusing stories about the unusual habits of this animal. According to one, some miners who kept a supply of nuts were astonished one day when they found that all their nuts had disappeared and instead there were a lot of useless stones in the box. The most amazing story is that of a gold prospector. One day he found, to his astonishment, a gold nugget on the desk in his cabin, which he believed had been put there by a packrat who had taken some of his objects in exchange. The man, the story says, followed the animal's trail, and as a result found a gold-bearing vein so rich in gold that it brought him great wealth.

Curiously, in the past, the Rannock Plain Indians used to show great respect towards this small animal, and in some regions it was commonly believed that it had miraculous power and could help humans solve their problems. It was customary to beg a packrat for assistance. When a man wanted success gambling he would go to the desert and, using special phrases, would appeal to the first packrat he met. He would even make a promise to the animal to do something in exchange if his wishes were granted, such as endure a long fast after his success.

In recent years the animal's peculiar habit of concealing theft by leaving other objects for the stolen ones has not been regarded as evidence of the animal's honest approach. The biologists who deprive the packrat of the romance of its habit say that when picking up an object he has to drop another one in its place because it is difficult for him to carry them all.

The packrat uses its stolen treasure to decorate its nest. The nest may be some 60 cm across and have several apartments such as a sleeping room, food storage room and even a separate room for rubbish. Each room is designed to serve a particular purpose for his family. The pack rat builds a nest which is an impenetrable castle. Using prickly cactuses the clever builder makes the entrance to the house so narrow that only an animal its own size can squeeze through without hurting itself on the spines.

Packrats live in the desert where there is no water to drink so they rely on the moisture content of fleshy cactuses. The cactus juice contains a large amount of oxalic acid which is very toxic to other animals, but is not hazardous to the packrat. The animal has specific enzymes in its body which metabolise and neutralise the toxins.

Packrats are monogamous and seem to enjoy the family life. But strangely, the males cannot always remain on good terms with their wives. Once the female becomes pregnant, she often loses her desire to share accommodation with her 'husband'. She may become so aggressive that she forces the male to leave the home. If he is reluctant to retreat she may bite him severely. Surprisingly, the 'gentleman' keeps his positive attitude and does not retaliate. He stoically accepts defeat without fighting and leaves the nest. He may come back, however, if his 'spouse' is in a better mood.

Packrats have another very unusual habit. When the mother and her babies are under attack, she runs away with all her litter hanging from her belly firmly attached to her teats. The teats must be especially adapted for such an emergency run, since the grasping and vigorous bouncing by the hanging babies cause no damage to the mother's teats.

11.
Mysterious Migrations

The Secret of Christmas Island

Christmas Island, which lies in the Indian Ocean some 360 km south of Java, is the land of crabs. They have found the tropical conditions of this island perfect and there are fifteen species of crab living in the forest covering the island.

The most common is the red crab (*Gecarcoidea natalis*) which lives in burrows on the forest floor. When breeding time comes, the whole crab population leave their burrows and march through the island to find their breeding grounds at the seashore. These crabs are so common that when a horde of them marches through the island, they literally cover large areas of the island with their red bodies. People are aware of this remarkable crab invasion and try to keep their doors and windows shut, otherwise the mass of crabs enter the houses. They can climb walls and fences and inside the house they enter every corner, including toilets and cupboards. During the invasion, which lasts about two weeks, all the roads are covered with a red carpet of crabs which make driving almost impossible as they puncture tyres with their claws. When the crabs cross railway lines many of them die from the heat as the sun makes the rails extremely hot.

It has been estimated that over 120 million red crabs participate in the invasion which usually commences in late October or early November every year, just when the rainy season comes. About one million of these crabs do not reach their destination because they are killed on the roads by passing vehicles or burned to death by the hot rails.

The habits of red crabs are very interesting. They live in the forest in their burrows feeding on leaf litter which accumulates on the ground. For several months they cannot be seen. All one can observe are their burrows plugged with dried leaves. An indication that there are crabs in the forest is an almost total lack of litter since millions of them can 'sweep' the floor thoroughly. Then at a specific time of the year the crabs leave their burrows and rush to the seashore which is several kilometres away. When they reach the sea they first bathe and drink as they are on the brink of complete exhaustion. Then the males retreat to the seashore where they establish their territory and make burrows. After mating, the females produce their eggs. Thousands of tiny eggs are tucked into the female's brood 'pouch' and then, after some time, the egg-laden females move to the shoreline where they release their young at night and then begin to crawl home.

The phenomenon of the red crab invasion is a great attraction to visitors to Christmas Island. But among several other species of crab living there one must be mentioned and this is the world's largest land crab called the robber crab or coconut crab (*Birgus latro*) which can weigh up to 4 kg. This crab is famous for its ability to climb coconut palm trees in its search for food. It breaks the young coconut with its powerful claws and eats the copra and drinks coconut milk which it relishes.

Some crabs are exceptionally large. The largest is the marine giant spider crab (*Macrocheira kaempferi*), so named because it resembles a gigantic spider. This crab is enormous and can have a claw span of 3 m. It occurs in the sea off the south-eastern coast of Japan. It is claimed that this crab may have a claw span of 5.8 m, but this has not been proved. The spider crab can be dangerous to humans because of its powerful claws which can inflict severe wounds and are said to have killed people in Japan. This crab is also vulnerable to predators, and, when first under attack, the crab tries to conceal its eyes by drawing them back into its head — a very unusual defensive ability. The smallest crabs are pea crabs (family Pinnotheridae), some species of which have a shell only some 6 mm across.

Divine Punishment

Probably no other animal in the world can cause people such misery as the insect called the locust. An invasion of these insects, which suddenly appear as if from nowhere and devour crops, causes disaster. In the Book of Exodus there is a vivid description of one such horrific locust attack which happened in about 1200 BC and devastated Egypt: 'They covered the face of the whole earth, so that the land was darkened . . . and there remained not any green thing in the trees, or in the herbs of the field, through the land of Egypt.' No wonder that some scholars blame the locust plague for changing the course of history, claiming that it caused such a famine that certain civilisations had to completely abandon their formerly fertile lands.

The invading swarms of locusts may consist of a hundred billion insects and can devour up to 20,000 tonnes of vegetation in one day. The swarms progress like a red avalanche destroying all plants on their way. All trees in the area are stripped down to naked bark and look like skeletons. All crops are destroyed and such an invasion usually leads to the starvation of both humans and animals living in the area. Since such a sudden appearance of an incredible mass of ferocious insects could not be rationally explained in the past, it was considered to be divine punishment for people's sins.

Locust invasions are most common in the warmer regions of the world. Karen Blixen, who witnessed such an invasion in East Africa, described it 'as a blizzard whistling and shrieking like a strong wind, little hard furious wings to all sides of you, and over your head, shining like thin blades of steel in the sun but quickly darkening the sun completely.'

The most dangerous is an invasion of the desert locust (*Schistocerca gregaria*) which has caused complete crop devastation in the past in many regions of Africa. The migrating swarms form such an incredible mass after landing that they can even stop a train. The wheels of a locomotive can become so clogged with the oily, crushed locust bodies that they are unable to move.

Ethiopians are among those nations of Africa which have probably suffered most from locust invasions in the past. One million people died of starvation when locusts devastated their crops during the First World War. People use magic to try to divert the horrible invasion of these insects. Shamans are invited to the fields which are likely to be invaded. In a loud voice the shaman repeats the phrase 'go, go, go' in the belief that this will divert the horrible plague. In one region people still believe that a swarm which devastated their neighbours' crops in 1958 did not come to their village because it was miraculously diverted by a shaman who altered its direction.

When a huge swarm of locusts invaded southern parts of Russia in 1879 the masses of insects covered the streets making them almost impassable. Locust bodies were almost everywhere; they even stuffed the chimneys of the houses to such an extent that bread could not be baked for several days.

People often try to protect their crops from an attack of locusts. When a locust swarm appeared in one part of Argentina, in other regions which were likely to be invaded, people in thousands rushed to their fields and erected barricades which were some 30,000 km long in order to stop the invasion but they did not achieve much. Even fire was used to try to stop the invading locusts. Despite the fact that many millions of insects became smothered, the remaining locusts marched forward over the charred bodies of their mates.

In some regions of Japan there is a traditional way of protecting crops from locusts. A phallus-like wooden figure is placed in the middle of the rice field in the strong belief that it will detract the locusts.

Strangely, a locust plague is not always a calamity and in some regions the locusts' arrival is a real blessing. In some sparsely inhabited areas of the Sahara desert, local people are happy when they see the swarms arriving because for them it means plenty of food will now be available. People are then busy collecting their harvest which has suddenly arrived from the sky. Many Arabs living in these regions regard roasted locusts as a great delicacy comparable to almonds in taste. Jews also occasionally had the chance to enjoy locust food. According to the Bible, John the Baptist survived in the desert because of locusts which were a miraculous 'gift' from heaven.

Locusts are grasshoppers that can be in either a solitary or a gregarious phase. Locusts in these two phases are so different that for a long time they were regarded as belonging to separate species. They differ not only in external appearance but also in behaviour, ecology and physiology. Moreover, there are also many intermediate stages. The change from solitary to gregarious phase is a very complex phenomenon and unfortunately it is still not fully understood. We do know, however, that it is the result of the degree of density of these insects, environmental factors and to a certain extent it is regulated by pheromones and hormones.

When the female locust is ready to lay her eggs she digs a hole in the earth by driving her abdomen into the ground with a rotating movement. During her digging her abdomen and some of her internal organs (such as the alimentary canal, nervous system and intersegmental muscles) stretch out to three times their normal size.

The Secret of a Butterfly

Among many species of butterflies the most famous is the monarch butterfly (*Danaus plexippus*) which occurs in the United States and southern Canada. This colourful but frail insect is among the most advanced of travellers as some of their number cover a distance of 3,000 km. Although it was known that the western population of these butterflies migrates to California, it was impossible for a long time to find the eastern population's winter destination. Despite intensive research, the site of migration of these insects remained unknown till about 1975.

After almost four decades of observations the problem was finally solved. We now know that the monarch butterfly migrates from eastern USA as far as Mexico where they land in the mountainous pine forests of Sierra Madre. The scientists who discovered them in this most remote region were astonished to see them in incredible masses covering every branch of the trees. It has been estimated that over 14 million butterflies can gather in just one location.

The monarch butterflies spend the winter in these high mountain peaks where the temperature is about freezing point. This seems to suit these insects well as they remain in a semihibernating condition for several months, keeping their energy reserves for the return flight. In this way the butterflies avoid the killing winter of their homeland and only make their long return trip home when spring comes. Unlike many other butterfly species that live for only a few days, monarch butterflies may live for up to nine months and this enables them to perform their two migrations.

On their return home, the butterflies mate and then lay their eggs.

They only choose one particular plant, namely the very poisonous milkweed, on which to lay their eggs. And there is a reason for such a peculiar choice. The butterfly caterpillars feed on the toxic leaves, but the poisonous substances do not cause them harm. On the contrary, they make them 'immune' to the attack of many birds which may be attracted to the juicy caterpillars. If a bird ate such an attractive meal it would become seriously ill because of the poisonous alkaloids in the caterpillar's body. It will learn a lesson, as do the other birds, and so most caterpillars survive and develop further.

This interesting adaptation has even further consequences. The adult butterflies which develop from the caterpillars also have the same plant poison in their bodies, located in the wings. In this remarkable way both the larvae and the adult monarchs obtain a strong weapon against predators. Of course, this protection does not work perfectly since many butterflies must die before the predator learns his lesson.

The presence of this strong poison in their bodies explains why these butterflies can congregate in huge masses when they land after a long migration and nevertheless are not decimated by birds and other predators.

Remarkably, some monarch butterflies occasionally migrate from North America as far as Australia. They sometimes reach Europe, but cannot breed there because of the lack of milkweed plants on which their larvae feed.

Some other butterflies which are somewhat similar in appearance to the monarch butterflies are also avoided by the 'careful' birds and thus have a high degree of protection. The monarch butterfly has become the symbol of California because of its beauty and its famous migration. In the town of Pacific Grove a monument has been erected in its honour and, in this region, farmers like this butterfly species because its caterpillars eat the noxious milkweed plant and keep it under control.

Equally fascinating is another butterfly species which is the world's largest. Called the queen Alexandra's birdwing (*Ornithoptera alexandrae*) this remarkable insect has a wing span reaching 28 cm, and is larger than many birds. These huge butterflies only occur in Papua New Guinea, and were once in such abundance that they were hunted with guns. Hunters used dust shot to bring the big insects down. They now fetch a high price as such specimens are sadly very rare. In an auction in France in 1966 one specimen fetched the sum of 10,000 francs.

The high demand for these once common butterflies and extensive hunting has led to a substantial decrease in their numbers. In order to increase the population, in some regions people started to grow plants eaten by their caterpillars. It is hoped that these butterflies could become an additional source of income for the poor forest inhabitants. Children sometimes use these butterlies as toys by attaching a line to them and making them fly like kites.

In ancient Greece butterflies were admired and believed to be the home of human souls and in ancient times in Germany there was a belief that all humans were reborn as butterflies. This fragile insect was even worshipped in some regions and considered to be the god who created the whole world. But in Serbia butterflies were considered to be the souls of witches and in some regions of Samoa people have a superstition that catching a butterfly can cause death.

Bird of Providence

Rarely have birds played such a decisive role in people's lives as have petrels known as Bill hill mutton birds (*Pterodroma solandri*) found on Norfolk Island. When, in 1890, the ship which regularly supplied the penal colony on this island did not arrive from New South Wales, famine gripped the settlers and only a miracle could have saved them from starvation. Suddenly thousands of mutton birds arrived to breed on the island. The birds landed in an enormous number — they 'were so thick that they came down a little after sunset like a shower of hail.' The hungry settlers had an easy harvest and killed the birds as fast as they landed. In one month over 50,000 birds were slaughtered and the colonists had plenty of delicious meat. Since the mutton birds had saved their lives the people named them the 'Birds of Providence'. Another name of the bird is Providence petrel.

One officer expressed his gratitude in the following words: 'The Mount Pitt birds, the birds bred in numbers on Mt Pitt, have been the greatest friends that ever any of us knew, for I may with truth say that they have saved all our lives, the greatest part of us would have been in our graves long ago if it had not been for these birds . . . I shall return my thanks for them the longest day I have to live.'

Most of the inhabitants of the island were convicts and, despite their gratitude, they continued to rely on these birds as their main food supply, even though there was a regular food supply by ship. They not only killed most of the birds, but they also collected their eggs and, in effect, the Norfolk Island bird population was soon completely exterminated. This species now breeds on Lord Howe Island.

Another bird, the short-tailed shear-water, which is also known as the Tasmanian muttonbird (*Puffinus tenuirostris*), breeds on the small islands of Bass Strait and Tasmania and used to occur in incredible numbers. Matthew Flinders was amazed by the incredible number of these birds he called petrels which passed over his ship in the Bass Strait: ' . . . during a full hour and a half this stream of petrels continued to pass without interruption, at a rate little inferior to the swiftness of the pigeon'. According to his estimate, the number of birds he witnessed then was 151 million. Although this figure seems to be an overestimation, it nevertheless gives an idea of how abundant these birds used to be in Australia.

Since Tasmanian muttonbirds are known for their delicious meat, it is not surprising that the Aborigines hunt them extensively. When the muttonbirds are coming to breed it is a happy time for the Bass Strait islanders. People even started to describe time as either before or after 'birding'. These muttonbirds are typical sea-birds and spend most of their time at sea. Although millions of them are spread over the immense Pacific, at a certain moment they start their long trip to the Australian coasts for breeding. The migration of some species is incredibly long as they may cover up to 30,000 kilometres in one season. Despite the fact that some birds are in remote areas of the sea while others are much closer to their destination, they all arrive at almost the same time. Incredibly, these wandering birds usually all arrive between 26 and 27 September every year and they even find the furrows they left the year before without difficulty. They arrive in many thousands and lay their eggs. But when the young hatch and mature, humans prevent many of them from reaching adulthood. The fat chicks are killed in the thousands. Oil is squeezed from their stomachs, feathers are used in sleeping bags and the meat is frozen and sold to restaurants.

Although naturalists regard the muttonbird harvest as a disgusting slaughter which should be stopped, some people consider it an amusing sport. Up to 4000 recreational 'birders' are registered in Tasmania who are said to catch over 300,000 birds each year. To prevent the bird's extinction, the harvest is controlled by a strict law.

The largest harvest of these birds takes place on Furneaux Islands in Bass Strait in Australia. Although people are allowed to hunt for only five weeks a year, the number of chicks collected during this short period reaches 700,000.

These birds were mainly hunted in the past for their oil which was commonly used for heating, but another application was also found for this oil. It was used in Australia for medicinal purposes. It was believed to be an excellent cure for rheumatism and tuberculosis and used to be sold in pharmacies throughout Australia.

As a result of extensive shooting and the intensive destruction of the muttonbird's habitat by farmers, their number has significantly decreased in recent years. If these unique birds are to survive, a long-term conservation policy is needed in Australia.

An Incredible Puzzle

Although people have always known of the European eel (*Anguilla anguilla*), the life history of this snake-like fish remained obscure for two millennia. Despite the fact that millions of freshwater eels had been caught and eaten around the world, nobody could find a young or baby eel. Neither had any of its eggs been found and nobody had ever seen the sexual organs of this fish. Small wonder that amusing stories have evolved about the origin of eels. According to Aristotle, baby eels were ejected from the 'entrails' of the earth. Many people believed that when horses' hairs fell into the river they turned into eels: some even thought that these elongated fish developed from earthworms.

Biologists were so convinced that these fish lacked sexual organs that when professor Mondini of Italy reported in 1777 that he had found an ovary in a female eel, he was simply ridiculed. His important discovery was ignored by scientists for another century until S. Syrski, a Polish biologist, found testes in a male eel.

Despite this progress, neither the eggs nor the young could be found until 1896. Two Italian biologists eventually discovered them in their own aquarium. It is quite an amusing story. These men had some peculiar, small creatures of a most unusual shape and colour in their aquarium. They were transparent and resembled willow leaves and were so amazingly different from any other fish that they were called 'ghostfish'. The two scientists noticed one day with great surprise that their two 'willow leaves' had disappeared, but found instead two eel-like fish swimming in the same aquarium. It was evident that the 'ghostfish', which had been considered to be an entirely different species of fish, were in fact the eels' larvae.

The eels' spawning grounds still remained unknown. Since the eel larvae, called leptocephali, were always found near the coast it was thought that the baby eels might have come from the sea. Solving the mystery of the eels' breeding ground became almost an obsession for Johannes Schmidt, a Danish ichthyologist, who spent many years studying these eels. He made many trips to the Atlantic Ocean and studied the young eel larvae which he caught in nets. He noticed that the further he was from land, the smaller were the eel larvae he caught.

When he reached the Sargasso Sea near Bermuda, some 4000 km from Europe, the leptocephali were only some 10 mm long and despite all efforts he could not find any smaller. So he came to the conclusion that this must be the spawning ground of the European eel. This was later confirmed by other scientists and the great mystery was finally solved.

It seemed incredible that the European eel lays its eggs in a sea 4000 km away from the European rivers in which it lives. What is even more surprising is that the adult eel's trip is its last as both parents die after fulfilling their parental duty.

The tiny larvae must also travel to their parents' home, covering the same enormous distance through the ocean. This trip takes some three years and during this remarkable journey the tiny larvae change dramatically, losing their willow-leaf-like structure. They become elongated, but are still transparent and this is why they are called 'glass eels' when they are about to reach European rivers.

Another interesting fact is that the Sargasso Sea is also the spawning ground of the similar American eel (*Anguilla rostrata*) and their leptocephali make a shorter journey, to the coasts of America, which is 'only' some 1500 km away. Although these leptocephali take only one year to reach America, when they arrive they look almost identical to the European glass eels. This indicates that the American leptocephali develop three times faster than the European ones.

Once the young eels reach the rivers or lakes of Europe or America they develop further and live there for several years. Then, one day, as if responding to a mysterious command, all the adult eels, which may be some 8–16 years old, commence their long trip to the spawning ground. They usually start their trip at night and they frantically swim downstream in an effort to reach the sea. Interestingly, those adult eels which live in isolated ponds also respond to nature's command, but they have to find the shortest route to the nearest river leading to the sea. They are able to wriggle like snakes and are known to cover considerable distances over wet grass until they find a river.

Before they enter the salty sea water, they change dramatically. Their digestive organs degenerate and from then on these fish are unable to feed in the sea. During their enormous trip across the ocean, they have to rely on the fat deposits accumulated in their thick bodies.

It should be noted that if an eel is kept in an aquarium and does not make the trip to the spawning ground where it would die, it may live for up to 60 years.

There are also some eels which live permanently in the sea. The most interesting is the so called garden eel (family Heterocongridae) which has most unusual habits. These eels, which are some 50 cm long, live at the sea bottom and the first biologists who saw them took them for grass stems moving in the sea currents. When hundreds of these eels stand in an upright position they form a kind of underground garden, hence the eel's common name.

It seems incredible that a fish would choose such an unusual way of living permanently in one place. The eels embed themselves with their tails in sand burrows so that the rest of their bodies rise up like walking sticks facing the current. These strange fish are very alert and can quickly vanish into their burrows when approached by a visitor. But after a while they all re-emerge and the 'garden' is soon revived as if nothing has happened.

Garden eels are so strongly attached to their burrows that it is impossible to dislodge them. Biologists who have wanted to catch a specimen have had real problems to overcome. Some have even applied the drastic method of using dynamite explosions to 'catch' this fish but soon discovered that a more gentle method can be used. These eels can be drugged. Marine biologists cover the burrow in which an eel is hiding with a kind of plexiglass dome and then introduce a liquid anaesthetic into the water under the cover. When the emerging eel swallows the 'poisoned' water, it soon falls asleep and can then easily be pulled from the sea bottom.

Since garden eels often live in the same place in large numbers only some 50 cm apart they attract many predators. This is why these eels come out of their burrows to feed for only a few hours a day and usually remain hidden in their safe burrows. But if they are not alert enough, a big fish such as a ray suddenly swimming over the garden eel colony can literally cut down all the eels like a lawn mower, and so garden eels are always on the alert.

12.
Incredible Civilisations

The Oldest Civilisation

There is only one creature in the world that can build a home one thousand times bigger than its own size. These animals are termites, erroneously called 'white ants', most species of which live in tropical regions of the world. The highest hills are constructed by *Macrotermes bellicosus*, common in East Africa, and can be over 8 m in height. The first explorers who found such hills in the African savanna thought that they had discovered an unknown human civilisation.

Some termites construct mounds for their millions of inhabitants in a peculiar shape. The compass termites (*Amitermes meridionalis*), found in northern Australia, build bizarre, wedge-shaped mounds, the flat sides of which face west and east. These strange up to 5 m-high hills make the countryside very peculiar. Surprisingly the way the tiny engineers construct their mounds indicates their 'knowledge' of weather conditions. The flat surface of the hill is only heated by the sun in the morning and evening, while at mid-day only the narrow edge faces the sun, and this prevents overheating. How these tiny, blind engineers can construct such a mound remains a mystery.

The termites make their homes from earth particles mixed with saliva. When dried the material is as tough as a brick. A huge elephant can even use it as a rubbing post and it will withstand the pressure. The tough termite hills, common in East African savannas, often cause problems. When a road is to be built all the hills have to be removed, but bulldozers are sometimes not strong enough and dynamite has to be used.

Termite mounds are also impressive structures when examined inside. Termites discovered, long before humans, how to maintain proper air circulation and introduced a simple but very efficient air-conditioning system maintaining suitable humidity and temperature conditions inside their 'home'.

To accommodate their different castes special chambers have been made by the tiny architects. Termites have well established castes, each specialised in performing various functions for the benefit of the whole colony. Unlike ants, the termite colony is a real 'monarchy' in which both the queen and king are 'in power'. The most important, however, is the queen who is constantly cared for by specialised workers.

The size of the termite queen is most surprising. She is so much larger than her 'citizens' that it is hard to believe that she belongs to the same species of insect. The queens of some species are up to 16 cm long and look like white sausages. Her abdomen expands 300 times its initial size and she is so obese that she can hardly move. She is, in fact, a most efficient egg producing factory, laying up to 36,000 eggs in one day. It has been estimated that a queen, who may live some 20 years, can produce over 260 million eggs during her life.

She is periodically fertilised by the king of the colony who is very small when compared to his huge 'wife'. It is interesting that the termite kingdom has no problem when the old termite queen dies. There is always a female who can take over as a new queen if necessary. These females are known as substitute queens. In a colony of *Amitermes laurensis*, native to Australia, there are about one hundred immature females which can become queens. According to some biologists, the clever termites ensure its colony's 'immortality' with such an arrangement.

People in Africa are aware of the termite queen's fantastic sexual reproductive capacity and believe that it can invigorate their own sexual power. Men collect the big queens from inside the hills and eat them as an aphrodisiac. They are also used as a remedy for various ailments. In some parts of Australia Aborigines use pieces of termite mound as a medicine for stomach disorders. We were quite surprised when one day an African came to our house, opened a box and offered us a huge termite queen which he had brought to sell us. Moreover in some parts of Africa people chew pieces of termite mound like chewing tobacco.

In some regions of South America, termite hills are used in peculiar rituals. There is a strange custom among the inhabitants of the Amazon river, that when a woman becomes a widow she cannot remarry until she goes through a purification ritual. She has to sit beside a chosen termite hill and both her head and the termite hill are covered with a cloth to form an umbrella. This is then set on fire which produces so much smoke that the poor woman is nearly suffocated. After this experience it is believed that the taboo attached to her is removed and she is free to look for another husband.

Termite colonies produce a huge number of reproductive individuals with wings and eyes who fly off from the termite hill in huge clouds in order to establish a new colony. The flying forms are white and when they fly in huge swarms one may think that it is actually snowing. Once, when travelling in West Africa, we were surprised to see 'snow' on such a hot night. The masses of white flying insects covered the car window to such an extent that driving became impossible.

These flying termites are both male and female. Soon after landing they frantically try to lose their wings by brushing their bodies against the ground. They then form sexual pairs and, after mating, look for a suitable hole to establish a new colony. But not many termites survive their nuptial flight, as most are devoured by lizards and birds. The local people also wait for such an occasion and gather them in large quantities, often with shovels and baskets. They are dried and then made into a delicious food. Many people who live near termite hills are so fond of the delicious termite 'meat' that they try to acquire an additional harvest by forcing the termites to 'escape' from the hill. They must have noticed that termites usually leave their nests just after rain so they make 'rain' by simulating its sound. To do this the natives surround the termite hill at night and beat their pots with wooden sticks to imitate rainfall. This is usually effective in fooling the termites which are ready to fly.

There are many species of termite, however, which do not construct mounds or termite hills but live entirely underground. Some even make nests on trees or on stumps. The subterranean termites are the most destructive. They build tunnels up the sides of the foundations of buildings and enter wooden structures eating the wood from the inside and only leaving a thin outer shell. There have been cases in tropical Africa when, during a dance in a private home, everyone has suddenly plunged through the collapsed wooden floor which had evidently been damaged by termites. Sometimes historical buildings have been lost to 'white ants'. A famous palace of the governor of Calcutta is said to have collapsed in 1814 because it was infested by termites. The tiny insects were once responsible for the fate of a battle. When British troops attacked a French garrison in the French Antilles in 1809 the first gun fired at the enemy caused the walls of the buildings to collapse, to the complete surprise of both defenders and attackers. The British were able to win the battle without much effort.

We were told of a story of a man in Kenya who blamed the 'white ants' for disrupting his wedding. He buried his paper money in his garden for safety, but just before the wedding he discovered that all the money had gone, evidently eaten by termites.

Termites have such an insatiable appetite that in their frantic search for wood they are known to eat various plastic insulations as if in the hope of finding edible wood inside. There is one voracious termite species in northern Australia which is particularly dangerous since it can gnaw its way through the lead covering of undergound telephone cables.

Certain termite species (genus *Macrotermes*) have not only mastered the art of making complicated mounds but also possess the knowedge of a trained gardener. These tiny creatures cultivate a special fungi. They carry out their agricultural work in special chambers and as the fungi grow they cut them down, keeping them small enough not to destroy their chambers. These fungi are particularly important to these termites as they help to break down the cellulose of the wood the termites eat. Unlike most other termite species, they do not possess symbiotic

Komodo Dragon

Thorny Devil

Frilled Lizard

Pufferfish

Perenty

Honey Ant

Superb Lyrebird

Koala

Red Kangaroo

Dugong

Nine-banded Armadillo

Satin Bower Bird

protozoans in their gut which normally fulfil this function.

Inhabitants of tropical regions are aware of the termite's gardening activities and they look for abandoned termite hills and collect the fungi which often grow in large numbers around old and destroyed termite mounds.

Peculiar Slavery

Ants are one of the most common animals in the world. Their number has been estimated in quadrillions, a figure impossible to imagine. Ants have always been admired for their remarkable social order and 'intelligence' and these small insects were even considered sacred in some cultures. Some Hindus still consider the tiny black ants as sacred and they feed them on specific days of the week to show their great respect. Hopi Indians in North America believed that the first people who appeared on earth were ants.

Ants have even played a curious role in some people's sex life. The strangest is perhaps the custom found on the Micronesian island of Ponape. A certain large and stinging ant is used to induce erotic sensation in women. The live ant is placed on the woman's clitoris and bites this highly sensitive part causing strong sexual excitement.

Among the Quabas people of South America there is a peculiar custom in which ants play an important role. During their initiation ceremonies they invite ants to check the young boys' readiness to become adults. The boys are ordered by the chief of the community to lay beside the nest inhabited by ferocious red ants and their bodies are smeared with honey to attract the sweet-loving ants. Soon the ants cover the boys bodies, but stuck in the thick liquid they cannot escape and start biting their skin. Although the pain is almost unbearable none of the boys complain. Passing such an ordeal is believed to change a boy into an adult and that from then on each boy will be changed into a man, immune to the misfortunes and pain of adult life.

The ant's 'addiction' to a sweet meal is so strong that some ants are ready to rob other insects of this delicacy. These ants choose aphids, also called greenflies, which secrete a sweet fluid called honeydew. The aphids feed on plant juices which they suck through their proboscis inserted into plants. The food on which the aphids feed is so rich that they cannot use most of it, and it is passed out again undigested, as honeydew.

The ants stimulate the aphids to release the honeydew from their anuses and lap it up eagerly.

For some ant species an occasional feast is not sufficient so they have invented a method which makes the tiny aphids available all the time for 'milking'. It seems impossible, but ants developed the very human institution of slavery long ago. When they find a large colony of aphids on a tree, they carry them all to a plant near their nest. Here the aphids feed on the plant juices and can be milked at will by the hungry ants.

Some ants even construct special shelters to protect their tiny 'slaves' from rain. The ants also protect their slaves from predators with their vicious mandibles.

Some ant species have applied another method. They invade an aphid colony at the end of the day and keep them in their nest all night. In the morning the ants escort their prisoners to a plant where they are then milked. There are even some ant species which collect aphid eggs and keep them in their nest for some time until the spring comes. These eggs are then carried onto a chosen plant to hatch. The ants wait until they finally develop into adults so that they can become their slaves to be milked at will.

Ants have many interesting habits and associations, among which is the association between the meat ant (*Irido myrmex*) and the imperial blue butterfly (*Jalmenus evagoras*) which lives in eastern Australia. The ants carefully observe the butterfly laying its eggs on the branch of a tree, and surround the eggs and guard them until they hatch. The emerging caterpillars are constantly guarded against predators until they begin to exude a honey-like liquid from their bodies. Only well developed caterpillars have the glands which produce the sweet honey for which the clever ants are waiting. To obtain the sweet liquid the ants stimulate the caterpillars by gently stroking their backs. Sometimes two or three ants 'milk' together walking on the caterpillar's back.

To obtain their 'sweet' meal the ants of a European species, *Lasius flavus*, called yellow ants, exploit the larvae of beetles of the *Lomechusa* species. The sweet liquid the ants obtain from these larvae is not just a delicious meal. It has a rather unusual effect on the ants which become hooked like humans after using hard drugs. The ants gradually become addicts of the larvae juice and establish a very strong bond with these creatures. They show such devotion to the source of their sweet meal that when under an attack they make every effort to save the beetle's larvae and leave their own offspring to the enemy.

The study of these ants has also revealed that the ants' addiction is not beneficial to them. Due to the toxic effects of the larvae juice, whole colonies of yellow ants have degenerated.

For some ants, however, 'keeping' slaves is a real necessity. The Amazon slave-making ants (*Polyergus rufescens*) are unable to feed themselves and would die if they did not have slaves. Their mandibles are modified into razor-sharp blades which are excellent tools, but only for fighting and killing. No wonder that these ants make a big effort to find colonies of certain ants of the *Formica* species. They defeat and rob them of their pupae, carrying them to their own nests where they grow into 'slave' workers. From then on they work for the benefit of their masters.

13.
Amazing Architects

Secret of Solomon's Wisdom

Beavers (*Castor* species) are among the largest rodents in the world and became famous for their beautiful pelts. But in the past these animals were mercilessly killed by men for another reason. The animals possess a scent gland called castoreum which became a wonder cure. It was in common use in ancient Greece as a cure for heart palpitation, hysteria and even as a cure for madness. As Sir Francis Bacon reported in 1605, 'castoreum destroys fleas, stops headaches, induces sleep and does much good to mad people.'

It was formerly believed that castoreum was located in the animal's testes, and that this animal avoided being killed when under attack by detaching its testes and so castrating itself. They believed that when a hunter got what he was looking for, he would leave the animal alone and it could escape. This strange habit was even reported by the ancient writers such as Pliny the Elder, and the belief in beavers' selfcastration was so commonly accepted that this animal became an Emblem of Prudent Sacrifice. It was believed in medieval times that the beaver should serve as an example to Christians how a prudent man must behave when attacked by the devil. He should throw all his sins in the face of the Devil and escape free of burden.

The beavers' strange behaviour has been recorded in the medieval books known as Bestiaries: 'The beaver, when it looks back and sees the hunter coming after it, at once tears off its testes with one bite and casts them before the hunter and fleeing thus escapes.'

Castoreum is actually located in two pairs of glands near their genitals and occurs in both sexes. Not only the beaver's castoreum scent was famous for its miraculous medicinal powers, its skin was believed to be even more important. According to a Jewish tradition, King Solomon's remarkable wisdom came from the beaver's skin. It was believed that he wore a beaver skin hat and thus acquired a remarkable memory and the power to properly judge events. According to the tradition 'to acquire a prodigious memory and never to forget what he had once read, it was only necessary to wear a hat of the beaver's skin.'

Beavers have always been admired by biologists for their unusual skills. They could build dams before man could imagine such structures. These animals seem to work according to a master plan. Their home is usually an elaborate lodge above high water level, made of sticks and stones plastered together with mud. To make the lodge safe against predators, the clever beavers construct it in the form of an island and the entrance is always below water level to make it invisible to predators.

Such a lodge once saved the life of an American traveller. In 1809, in the region known today as the Yellowstone National Park, a man was chased by a group of unfriendly local Indians who wanted to kill him. The Indians were astonished when the man dived into a pond and vanished. Actually, after diving in, the man crawled inside the large entrance of a beavers' home and found sufficient room to hide. Since the floor of the beavers' home is above water level, the escapee could remain there for a long time until the puzzled Indians went away empty handed.

The clever beavers seem to know that during the spring floods their elevated 'home' could be flooded, and to prevent this happening they construct a dam. They are experts in felling trees which provide the building material for their dams. Using their powerful teeth they can cut down a tree of 10 cm diameter in just fifteen minutes. They work hard and are even able to cut through a trunk some 0.5 m thick. If a trunk is too long to transport to the dam site they cut it into shorter pieces. The beavers float their logs on the water to move them and sometimes build canals to do the job. Having transported the material to the dam site, the animals place the logs layer on layer until the proper dam height is obtained. The dam has a spillway to keep the water level constant near their lodge.

Beavers work extremely hard and some dams constructed by these animals may contain one hundred tonnes of material. The largest dam constructed by beavers was on Jefferson River in Montana and was 700 m long. The animals cut an enormous number of trees for their constructions. They usually work in pairs and one pair of beavers in a Canadian national park was reported to have cut down 266 trees to construct three dams about 15 m long across a stream. Large dams are the result of the work of many beavers. Some dams are used by several generations and are always maintained in good working condition.

There are two species of beaver, the European beaver (*Castor fiber*) and a similar but larger American beaver (*Castor canadensis*) which became a symbol of Canada. The European beaver was once very common, but has almost been exterminated by men greedy for castoreum. In Canada beavers were slaughtered for their pelts. The demand for beaver fur was so high that between 1853 and 1877, three million beavers were killed. Although the animal is admired in Canada and the first Canadian postage stamp issued in 1851 was the three-penny beaver, the persecution of this national emblem was horrendous. Fortunately, beavers are now protected and a great effort is made in Canada and elsewhere to reintroduce them.

The World's Finest Thread

The remarkable skill of spiders to produce mysterious webs has been admired since ancient times. These rather unattractive animals were even venerated in some cultures. The Bhils and Mats of India regarded certain spiders as their ancestral spirits and worshipped them. Spiders are admired in certain regions because they are believed to have helped people in danger.

There is a story that a spider saved the life of the great prophet Mohammed. The spider built a beautiful web across the cave in which the prophet had hidden in an effort to escape from his persecutors. The attackers carefully checked every cave except this one. Having found an immaculate web at its entrance, they were convinced that nobody could have passed through.

Although the spider's web looks so delicate its strength is remarkable. In some species it is stronger than steel thread of the same thickness. Some spider threads are only about one micron in diameter and finer than the thinnest man-made thread which is produced from gold and is about four microns in diameter.

The strongest web is produced by the spiders of the *Nephila* species and in some Pacific Islands people use the web for a rather unusual purpose. It is used to punish unfaithful women. The victim's head is covered with spider's web which causes death by suffocation.

Other spiders' web is so strong that it can be used as a 'net' for catching fish. The Australian Aborigines are experts at using spiders and their webs for catching fish. They first insert a stick covered with slime into a spider's web to 'catch' the spider. A web is attached to the end of the fishing line instead of a hook, while the spider's body is thrown into the water to act as a bait. The fish approaches the bait and becomes entangled in the web and is then pulled out of the water. An easy catch indeed.

Scientists became interested in spider's webs and examined the master builders' skill when under the influence of certain drugs. The spiders reacted immediately and the web they produced changed dramatically depending on the substances applied. When a spider was fed with a caffeine solution he produced an abnormal web which did not resemble a 'normal' web structure. Strangely, when a spider was fed with a solution of LSD this hallucinogen caused the spider to create a much better web than usual.

Some spiders produce gigantic webs, long enough to pass over a small river. There are also species whose web is no bigger than a normal sized postage stamp. With such a minute web at its disposal, the spider catches insects by holding the small web between its front legs like a man catching butterflies in a net and it may be said to use a tool to catch food.

Although it is generally assumed that all spiders make webs of various sizes and patterns, some species do not make webs at all but only produce long strands of silk. The threads end with sticky globules. The clever hunter throws its line like a man using a lasso. As a result an insect is usually caught by the line and the spider pulls it to its mouth and kills it with a poisonous bite. These peculiar spiders are called bolas or angler spiders.

As mentioned, the strongest web is created by the giant wood spider (*Nephila maculata*), which is among the world's largest species. Native to Brazil this spider's total span across the legs is 20 cm and it was believed that these spiders even overpower and devour birds. These spider stories were not believed. In 1705, when Maria Sebilla Merian published the first report of a giant *Nephila* spider sucking the blood of a hummingbird, she was ridiculed by biologists. It was not believed that a spider could devour a bird and her story was not accepted as true for over 150 years. It was in 1863 when H. W. Bates witnessed a similar event in the Amazon Valley that the existence of the remarkable spider 'beast' was accepted in scientific circles. He saw a *Nephila* giant spider busy eating a finch, which it had captured in its strong web.

Although people are afraid of spiders as some can inflict a deadly bite, sometimes an almost innocent spider can be blamed for causing dangerous symptoms. Strangely, one of the tarantula spiders (*Lycosa narboensis*) found in southern Italy was blamed for causing a terrible bite which was said to produce intense pain and even delirium. The only effective 'remedy' was vigorous dancing accompanied by special music for the occasion. This belief persisted till the second half of the nineteenth century. Some historians believe that the 'sick' dancers were not ill at all, but used the spider as an excuse for performing certain pagan ritual dances which were persecuted by the Christian authorities. Even today a very vigorous Italian folk dance is called the tarantella.

It is interesting to note that there are some sociable spiders and some colonies may contain as many as 100,000 individuals and make and inhabit huge communal webs. These spiders are known not only to live together in one location but to work together when capturing their prey and even provide communal care for their offspring.

'Dog Towns'

Despite their common name, prairie dogs are not dogs but short-tailed ground squirrels (genus *Cynomys*), which were so named because of their strange barking call. The first settlers in the USA called these animals the wild dogs of the prairies. In some parts of Texas prairie dogs were so common that towards the end of the nineteenth century the total population in this state was estimated to be some five billion. One of the largest prairie dog 'towns' which covered an area of about 400 km long and 160 km wide, was once inhabited by some 400 million animals — the largest town in the world.

Since these animals live in deep burrows, large areas of fertile land in the USA were useless because the whole area was covered with holes. Moreover, because the tunnels are often situated only a few centimetres beneath the earth's surface they often became traps for galloping horses which broke their legs when stepping into these hidden tunnels.

Early settlers hated the prairie dogs and made every effort to kill as many as possible, in order to make the land usable. Nobody admired the unusual engineering skills of these animals although in this respect these rodents are truly remarkable. They build underground 'towns' which consist of kilometres of well-made tunnels and dens extending in every direction beneath the soil.

The entrance hole to their 'towns', which may be only some 8 cm across, goes almost vertically down for some 4 m before it levels off to a horizontal subway which joins the large nesting chambers and other compartments. The animals have special chambers where they sleep, or care for their babies. Moreover, the clever animals make turning bays in their underground labyrinth, where they can hide from predators. They deliberately construct the entrance to the tunnels in a mound, which is an excellent protection from floods. It is also a place where their sentinels can watch and warn their mates. They stand on guard like soldiers, and when they spot a predator they give two sharp barks as a warning. These calls are well understood by all the colony members who escape and hide in their burrows. When the danger is over the guards give another loud message to the colony. Although many thousands of prairie dogs live together in complicated burrows they never miss their own den.

Prairie dogs occasionally share their homes with deadly rattlesnakes. It is still unclear why they allow this or why they sometimes even deliberately extend their burrows to make the larger snakes a more comfortable shelter, even though the visiting snakes are known to consume their youngsters in the breeding chambers. Some biologists claim, however, that the prairie dogs do not like their 'guests' and that they sometimes try to seal off big rattlesnakes in their tunnel in order to bury them alive.

Prairie dogs seem to exhibit a kind of human-like behaviour. For example, when two females meet outside the burrows, they greet each other by standing up with their forepaws touching, and touch each other's mouths as if kissing. A male and a female also 'greet' each other with appeasement gestures when they meet. But when two old males meet they are not so friendly and whirl around, trying to cover each other with kicked up dirt.

The prairie dogs work hard to keep their town in good order and repair their chambers when they are damaged. They even bury their dead, covering them with soil.

Unfortunately, these interesting animals make the land in which they construct their tunnels unsuitable for farmers who consider them a pest and try to eradicate them completely. Millions of prairie dogs have been killed.

Since 1900, the prairie dog population in the USA has been reduced in many regions by more than 90 per cent. Some conservationists blame the name of the animal for its fate. They argue that if these animals had not been called 'dogs' they would have become a fine game animal. As a result, they would have been controlled and their development would have been encouraged in some areas. The flesh of the prairie dog is superior to that of the wild rabbit, but because the animal was called a 'dog' the early pioneers refused even to touch the meat, even though it was not a dog at all.

Strange Farmers

People who visit tropical America are often surprised to see an unusual procession of ants walking along the road and carrying large pieces of leaf above their heads. It looks as if these ants have covered their bodies with parasols, hence their common name, parasol ants. The most common are the parasol ants belonging to the genus *Atta.* They climb a chosen tree in many thousands and then each ant cuts out a fragment of leaf by biting an arc from one side of the leaf to the other as far as they can reach. A detached piece of leaf is then carried above each ant's head as they return to the nest. A colony of ants may cut off all the leaves from an entire tree during a single night. For this reason they are also called leaf-cutters.

Close observation has revealed a strange fact. On the top of each leaf piece there was always a passenger, a small ant which seemed only to be there for a free ride. This strange behaviour puzzled scientists for years. It has finally been revealed that the ants do carry these small ants, and that the tiny passengers are in fact small soldiers who sit on top of the leaf to protect their mate from another insect predator. The *Atta* ants are most vulnerable to an insect species which lays its eggs on their necks and can kill the ant when it hatches. Hence the necessity to increase their load by inviting a guard on top of the leaf.

It was formerly believed that leaf-cutting ants simply use leaves for food. Thus scientists were truly surprised when, in 1874, Thomas Belt discovered the unique farming skill of these ants in Nicaragua. He discovered that the ants chew the leaves, mix them with saliva and their droppings and make their own compost which is placed in special underground chambers with the mycelium taken from the fungus growing in an exhausted bed.

These tiny farmers take care of their farms and periodically kill other unwanted fungi which may grow by injecting them with a special secretion produced in their own bodies. Thus, the clever ants behave like modern farmers who apply fungicides on their plantations.

As the fungi's fruiting bodies may grow too large and crush the chamber where they grow, the tiny ants control its growth. They systematically bite off the ends of the small fungi and eat them. A single nest may have as many as one hundred fungus gardens to supply a colony of millions of individuals. The nest of an *Atta* ant colony may be located up to 5 m below the ground.

Interestingly, each species of fungi-growing ant uses a specific kind of fungus and for each colony its particular mycelium is its most precious possession, since without it the whole colony would perish. When the time comes for a new colony to be formed, the new queen always takes a small portion of the mycelium, carrying it in a special recess inside her mouth. Once the colony is fully established and the growing chambers are ready, the queen passes the precious mycelium to her workers.

14.
Tool-using Creatures

Forest Doctor

Although apes and some monkeys, the only animals with 'human' faces, should be admired, in the past these animals were considered to be caricatures of man and were hated. Regarded as horrible brutes, they were related to evil, so that to dream of a monkey was a sign of misfortune. Even in the official view of the Church in Europe, monkeys were diabolical beasts which could do nothing but imitate people and were called monkey-devils.

A strange method of punishment was connected with these animals. When a man was accused of a crime against his own father, he was placed in a sack with a live monkey and drowned in the river or sea. But the traditional concept of the monkey as a fool and a sinner declined in nineteenth century Europe when apes were investigated. Scientists were fascinated by ape behaviour and they started to argue that the ape's intelligence could be compared to that of humans, an idea which shocked many religious people and provoked intense dispute.

Among the apes which have the highest degree of intelligence is the chimpanzee (*Pan* species). This animal, which is found in tropical Africa, can be tamed and trained to perform many things just like humans. Chimpanzees soon became circus artists and amazed crowds. There is even one account of a trained chimpanzee in America who used to drive a car down the town's busy streets. A surprised policeman saw a chimp at the wheel of a speeding car. At his side was the owner of the car who was busy changing the gears. He was forced to pay a speeding fine.

Chimpanzees have learned to use such tools as a hammer or keys, they have been taught to put clothes on and they can even be seen eating at table where they often behave better than some humans. No wonder that a man once employed chimpanzees in his factory where they learned to stuff foam rubber into cushions. His only complaint was that his 'workers', who worked extremely fast, often refused to continue when their mood suddenly changed. A chimpanzee once 'worked' in a restaurant. When asked for a glass of soft drink, this unusual servant took a bottle, removed the cap, put two straws in it and gently handed the drink to his customer.

Some chimpanzees have even become painters. The pictures produced by one chimpanzee 'artist' were once exhibited in a London gallery and the chimp's owner was most delighted with his pet's talent.

Chimpanzees can learn to communicate using American sign language using as many as 150 signs. A chimpanzee can even invent his own signs. As one author noted: 'one chimpanzee . . . learned to give the sign for open when she wanted to go through the door and amazingly, used it when she wanted a water tap turned on. She also coined the phrase "open-food-drink" for a refrigerator, and when she heard a dog bark, she said "hear-dog". The main problem for the chimpanzee is to reproduce the sounds of human words. Their best achievement so far has been to pronounce three words, namely mama, papa and cup.

At the early stage, the intelligence levels of a chimpanzee and a human child are very similar. According to Reynolds: 'Chimpanzee (named) Gua was tested in comparison with the boy Donald at 16 months. She passed 15 out of the Gesell tests (for preschool children) while Donald at $15^1/_2$ months passed 16.'

Untrained chimpanzees living in the wild also use tools. They crush leaves, for instance, to make a sponge which they use for drinking water or for removing dirt from their bodies. To catch termites, which the animal relishes, the chimp finds a twig of the proper thickness and size, removes the leaves and even pares the end with its teeth so that it penetrates the tough termite mound more easily. This tool is inserted into the termite nest and when it is laden with insects it is withdrawn and the insects are eaten. If a banana is too high for a chimp to reach, he looks for a stick long enough to get at the fruit. If a long enough stick is not found, he may join two or more hollow sticks together to make them into an efficient tool. By contrast, a gorilla would never grasp the idea even if a man showed him how to do it.

Chimpanzees nurse each other in an emergency. This ape, for instance, can remove an irritating foreign body from the eye of its suffering mate. Like a human, it first inspects the eye and then, pulling down one lid, gently removes the grit using its fingers. An astonishing first aid performance!

Even more amazing is the fact that chimpanzees are their own doctors. It has recently been revealed that chimpanzees living in the African forest use herbs in an evident effort to cure their ailments. A sick animal was observed searching for a particular herb in the forest which he then swallowed. These animals seem to be good herbalists as they know how to choose a medicinally active plant from among thousands of plants growing in the forest. It has been found that chimpanzees use exactly the same plants that are favourite herbal remedies among people living in the same region. When a local man has a stomach ache he searches for a certain plant known as Lippia, which is also a popular remedy used by chimpanzees.

Chimpanzees use some plants, however, which have not been used by the natives. When these plants were investigated, the chimpanzees were found again to be right, since these plants had a number of medicinally active ingredients. One of the plants 'discovered' by the chimpanzees was *Rubia cordifolia*, found in Ugandan forests, which was found to contain, besides other medicinally active ingredients, some active anticancer compounds. Scientists believe that by 'imitating' the chimpanzees' jungle pharmacopiea man will learn from apes how to cure some of his diseases. How the animals can find the herb they need among thousands of plants remains a mystery.

These intelligent animals can become real friends of humans and in some cases their owners' devotion is amazing. One chimpanzee became a great friend of Michael Jackson, the famous American singer. This beautifully dressed chimp is known to travel with the artist and eat with him at the same table and is said to behave like a real gentleman.

The chimpanzee male has testes which are approximately three times larger than those of a man. As he is a promiscuous animal and mates very frequently the testes have to produce a large amount of sperm.

The Bird of Death and Resurrection

The vulture's survival depends on death. These carrion-eating birds search for the dead bodies of other animals. There used to be a common belief that wherever a swarm of vultures arrived there would certainly be a human death. These birds were said to follow armies and to position themselves before the battle between the front lines, waiting for a 'good harvest' to come. Some commanders used to forecast the number of soldiers which would die in the battle by counting the number of vultures circling around. During the Crimean War in 1854, so many vultures arrived on the battlefield that they not only devoured the dead but even attacked wounded soldiers. To prevent this from occurring special shooting squads had to be posted after the battle was over.

Although this bird has become a symbol of death, in some regions it has been linked with resurrection. The Parsees of India, who have retained certain rituals of ancient Zoroastrianism, invite vultures to help them in their peculiar funerals. These people place the naked bodies of their dead on elevated structures called 'towers of silence,' expressly to lure vultures. The birds circling in the air immediately detect the opportunity and arrive in large swarms. They frantically devour the human flesh, leaving only the bones. The sect members do this because they believe that the faster the remains of the human body disappear, the faster the man will be resurrected in the afterworld.

Vultures also played a significant role in the religion of ancient Egypt. The Egyptian goddess of eternity, called Maut, was portrayed with a vulture's head.

As it is normally difficult for a vulture to find carrion, it spends a lot of time circling high in the air observing a wide area to locate a recent death. The birds also carefully watch the behaviour of their mates as many gathering in one place indicates the presence of a carcass. A hundred of these birds may soon arrive from a distance of up to 60 km to take part in a feast. They literally cover the whole carcass and may gorge themselves so much that they are unable to fly.

Although vultures may be despised for their voracity, they are in fact very beneficial to humans in tropical and subtropical regions as they clean rotting carcasses and help to keep the area sanitary.

Some vultures show a high degree of intelligence and for this reason are admired by biologists. The Egyptian vulture (*Neophron percnopterus*) is among the few animals which can use tools. This bird particularly relishes ostrich eggs, but since it cannot break their hard shells with its beak, the bird 'invented' a special technique. It uses a stone to break the shell. When it finds an ostrich egg, it looks for a stone of a suitable size. Manoeuvring in the air, the bird lands in the right position and then starts 'bombarding' the giant egg with the stone, throwing it with its beak. He may hit the target several times until the egg breaks. Sometimes the bird is unsuccessful, and may look for a bigger stone.

It was thought initially that the tool-using skill was inherited but this is not true. The young vulture has to learn. The stone throwing skills of the young birds vary greatly. While one young vulture may be unsuccessful after spending 30 minutes trying to hit the egg, another may arrive and crush the same egg with his first hit. He is clearly better 'trained' for the job than his mate.

When an adult vulture finds an ostrich egg, he considers it to be his own property and drives away other birds attracted to it. The other birds are evidently angry and express their mood by throwing their stones to the ground.

These intelligent birds do not use their tools when dealing with smaller eggs. The vulture simply smashes such an egg by dropping it onto a rock from the air. A few other unrelated birds are also known to use tools. One such bird is the Australian mudlark (*Grallina cyanoleuca*) which behaves like a real nurse. When a bird has broken its leg another will immediately pull out some of its feathers, dip them in mud and then plaster the injured leg. It is genuine first aid as the splint is applied correctly. It remains a mystery how the uninjured bird 'knew' that the mud would stiffen and help to firmly hold the two parts of the broken leg to facilitate healing. Some birds are known to use tools for fishing. For example, the squacco heron (*Ardeola ralloides*), found in the Mediterranean, catches insects, but rather than eating them it uses them as bait. The bird throws the insect into the water and when a fish comes to catch the bait, the watching bird immediately kills the fish with its beak.

The woodpecker finch (*Carmarhynchus pallidus*) of Galapagos uses a long, sharp cactus spine as a tool. In order to catch insect larvae he probes the bark of a tree with this sharp instrument. Without this tool the bird would be unable to get at these delicacies hidden in the tree.

Living Jar

Certain ants found in desert regions have evolved the most unusual method of food storage. Unlike bees, which store their honey in waxen cells, these ants store their honeydew in the bodies of their own mates. These strange ants, which occur in some regions of Australia, New Guinea, USA and Mexico, are called honeypot ants.

After the worker ants collect the honeydew they feed the sweet fluid to certain selected workers. Each ant has a food storing crop which is filled with such a colossal amount of liquid that it swells to the size of a large pea. Such an ant, which is called a replete, is unable to move and is suspended from the ceiling of a special chamber. When many repletes are suspended they resemble minute amber bulbs hanging from a ceiling.

This unusual food store helps these ants to survive during periods of severe drought when food is scarce. The ant workers then regularly visit their strange restaurants for a sweet meal. They gently stroke the 'honeypots' and by keeping their mouths close to those of the repletes they make them regurgitate the sweet honey from the storage 'vessels'. Some 'honey' becomes food for visiting worker ants, but much of it is distributed to other members of the colony and is especially useful for feeding the larvae.

It is still unclear how long these repletes hang in total darkness in their chambers. Neither is it known what happens to them when they become empty.

Not only ants have the pleasure of consuming the 'honey pots' sweet meal. People living in desert regions have also discovered this source of sweet delicacy. Small wonder that the Australian Aborigines have been familiar with these ants' strange habits since time immemorial. In fact, Aboriginal women spend much time searching for the sweet, underground meal.

These women are real experts in detecting these particular ants' nests and they know how to excavate the right underground chambers to find honey pots. They dig huge holes in the sand, often using only wooden sticks and they work with astonishing speed. According to Sir Baldwin Spencer, who observed them in action, '. . . acre after acre of hard sandy soil is seen to have been dug out, simply by the picks of the women in search for the insect, until the place has the appearance of a deserted mining field where diggers have long been at work "prospecting".'

Aboriginal men share the meal when it is found. Although they never participate in digging for honeypots they are the first to enjoy the sweet meal as the women present them with their harvest. Placing the repletes in their mouths, they bite off and spit out the upper part of the ants' bodies while honey from the 'bulbs' flows into their mouths. Special ceremonies are performed in some regions by the Aborigines to assure future good honeypot harvests.

The honey ants native to Australia (such as *Melophorus bagoti*) are most common in Central Australia. The Papunya people believe that the honey ants are their ancestral spirits and the name Papunya actually means 'Honey Ant Dreaming' and is related to various myths. According to one, where these ants now live there used to be ancestral honey ant people.

The honey ants which occur in the USA and Mexico belong to the genus *Myrmecocystis* and in the chambers of one big colony there may be up to 2,000 repletes which are often collected by native people. They are even sold in local markets in Mexico and are regarded as a great delicacy used on special occasions. In some regions of Mexico these unique 'sweets' are only served during wedding ceremonies on silver plates.

A Living Tool

It sounds incredible, but certain ants know how to use tools. Such are the tailor or weaver ants. These ants, which occur in most parts of the world, make their nests from leaves. They do not cut them off, but use them still attached to a tree. To construct their chamber-like nest, the tailor ants work according to a plan. They first select two leaves which are closest to each other and their main problem is to bring the leaf edges together. The ants grasp the edge of the first leaf by holding it with claws on their hind legs. Then they use their mandibles to seize the edge of the opposite leaf, so forming a living bridge along the full length of the two leaves. Sometimes, however, the ants cannot form a complete bridge along the entire length of the leaves because in one place the gap is too wide and as a result some ants are left hanging in space. In such a situation other ants help. They are seized at the waist by the hanging ants and these 'extra' ants are able to grip the distant edge of the leaf in their mandibles. If the gap is still too wide, a living bridge of up to four ants is formed.

If the leaves are ragged, another team of ants arrives and trims the edges. This is done before the second leaf is seized. Once the two leaf edges are held firmly the ants pull them together. When this is accomplished the leaf edges have to be stuck together, but the ants lack glue and so the ants have made a peculiar 'discovery'. They have found that their own larvae can be used. Another team of worker ants arrives, each carrying a larva between its jaws. By pressing the larvae hard against the leaf, viscous silk is secreted from the grub like paste from a tube of toothpaste. By moving this peculiar living tool back and forth, the worker acts like a man using a shuttle in a cotton factory. As the silky thread hardens, the edges of the two leaves are strongly glued together. By using their own larvae the ants can successfully complete their shelter.

The larvae of these tailor ants are believed to show their willingness to serve as living tools by sending signals to the worker ants. As the larvae's silk is used for nest building the larvae cannot make cocoons.

Early naturalists, who were not aware of the ant larva's ability to produce silk, believed that the ants invited spiders to make silk for them when making their nests.

The larvae are also needed for repair work when the nest chamber is damaged. The tailor ants strongly defend their nests and use their vicious bite against intruders. They guard their 'territory' so well that no other ant species ever 'dares' to try to make their own nests on the same tree.

The tailor ant's ability to deter other ant species has even been exploited by humans in the past. The early Chinese, whose citrus trees were attacked by crop damaging ants 'invited' tailor ants to deter them. They collected weaver ants with parts of their nests, covered them with gauze to prevent them from escaping and hung them on the citrus trees. At a time when insecticides were unknown the ants were of great value to Chinese gardeners. There were even special tailor ant collectors who 'prepared' them for the market and made good profits from such trade.

Among the best-known tailor ants is the African weaver ant (*Oecophylla longinoda*) which occurs in the forests of tropical Africa. As its colonies are very large, some composed of over 100,000 members, they construct a large number of nests. There could be up to a hundred nests hanging from the branches of one tree and such a tree, seen from a distance, looks quite extraordinary.

Three-Humped Monster

The sea otter (*Enhydra lutris*) is among the few mammals which deliberately use tools. Although their tools are simple stones, the way they use them signifies a high level of intelligence. When the animal dives and finds a clam which is difficult to open with its teeth, it looks for a suitable flat stone. With the clam and the stone the animal rises to the sea surface. Floating comfortably on its back, it places the stone on its chest as a sort of anvil. Holding the clam in its 'hands', the animal starts hammering it against the stone. It is usually hard work and takes many blows before the clam breaks open to expose the juicy flesh.

Sea otters are often known to use the same 'tool' again and again. They even hold the stone when diving in search of more food and this in fact increases the diver's total weight and facilitates diving.

The sea otter also acts intelligently in another situation. Sea urchins sting when handled because they are covered with spines and when the clever sea otter finds one, it first wraps its prey in seaweed to break off the spines before eating it.

These animals live off the Pacific coast of North America and the Bering Sea. They spend their entire lives at sea and to obtain their food they can dive to depths of up to 40 m. Sea otters have a strange habit of floating on their backs. They even sleep in this position and to prevent being swept away by sea currents, the clever animals entwine themselves in seaweed. These social animals are known to live in large groups of up to a hundred members.

The sea otter female is a devoted mother and is busy nursing her baby for several months. The mother suckles and cares for her baby by holding it on her chest while floating on her back. She often plays with her baby and is even known to toss it in the air and then catch it. When the female dives for food she hides her baby in the seaweed. When attacked by a predator while holding her baby, a sea otter female will defend her baby and fight till her death, but never abandons her child. If the child is grabbed from her, she cries loudly like a small baby.

The sea otter remained unknown to the western world till the first half of the eighteenth century. It was discovered by Vitus Bering while on an exploratory sea voyage on behalf of the Russian authorities. His ship struck a rock and was seriously damaged and the crew took refuge on one of the desolate islands. They were starving until they found vast numbers of an animal they had never seen before and they had plenty of meat to eat. These were sea otters, and the sailors also soon noticed that they had beautiful fur. Finally the crewmen managed to repair their vessel and return to Kamchatka, taking many sea otter skins with them. The new furs caused such a great sensation among traders that soon special expeditions were organised to collect the furs. Russians could make great profits selling the new furs to Chinese merchants as they were in great demand among the Chinese aristocracy.

The demand for sea otter fur was so big that Russians forced the Aleut Indians to work for them, hunting the precious sea otters. The Indians became expert hunters. They chased them in small canoes usually disguised with sea otter skins to fool the animals. If an Aleut returned home empty-handed, he believed that he had been unsuccessful because his wife had been unfaithful. Some similar otter species are found in fresh waters. They inhabit rivers and lakes of most regions except Australasia, Madagascar and the polar regions. Many otters live in African waters.

An amusing event is related to these animals. Once when Theodore Roosevelt was in a boat in the large Naivasha Lake in Kenya, he suddenly saw a peculiar three-humped monster moving across the lake. He had been told before that a mysterious three-humped monster lived in this lake which could occasionally be seen on the lake surface. The excited traveller took aim at the 'monster's' central hump and fired his gun. Surprisingly, two of the humps disappeared while the third, the central one which was hit, remained floating on the surface. When he and the accompanying men came closer they found one dead otter. They realised that what they had seen and what had been taken by many observers to be a mysterious three-humped monster living in the lake was three large otters swimming in a line. This event became so famous that the skin of this African animal can still be seen at the American Museum of Natural History.

15.
Implausible Skills

The Headless Learner

At first sight it seems that there is nothing interesting about the 'filthy' cockroach. Yet they are one of the most amazing creatures in the world though many of their habits and abilities remain unknown to most people.

Probably most incredible is the ability of some cockroaches to live for some time after their heads have been cut off. The cockroach from the suborder Blattaria can not only remain alive without its head for up to seven days, but, incredibly, it is still able to learn. It has been demonstrated that such a cockroach can learn to lift its leg out of water whenever an electric shock passes through it to the insect, and this indicates that it has another learning centre which works independently from the head ganglion.

Cockroaches are fantastically adaptable to living conditions and can survive without food for several months. Even after being frozen in ice, once it is thawed, the insect will 'wake up' and walk as if nothing has happened. Their ability to survive is incredible. It was found that cockroaches can withstand a level of radiation 100 times higher than that which kills humans. Some scientists say that if atom bombs destroyed all living creatures on our planet cockroaches alone would have a chance to survive.

These insects can be found almost everywhere, even in hot deserts such as the Sahara. They are found in the bottom of streams, and some live inside ant nests where they feed on the ant fungus gardens. These clever insects are even known to hitchhike. They sit on the backs of certain queen ants and fly with them during the mating flight.

Cockroaches seem to eat anything. They may feed on glue, wax, soap, or other products which are not normally regarded as food. They can sip beer from a bottle and they are even known to devour human hair or fingernail clippings. When very hungry they eat the eggs laid by their own females.

Cockroaches can be found comfortably living in the most unpredictable places, because they can eat whatever is available. An East African cockroach (*Supella supellectilium*) was found living inside a TV set where it fed on glue used in the insulation. It must have found sufficient food there as it never left the shelter. Biologists have wondered how this cockroach could survive so long without water and some speculate that the animal must produce water molecules from its solid food components.

Although cockroaches possess wings and some occasionally fly they prefer to use their legs. They are efficient runners, and have a remarkable ability to fool a predator and escape in time. The cockroach has a very sensitive device in its legs and reacts to the slightest earth tremor made by a predator or a walking man. Its response to danger is extremely rapid. In fact the impulses which travel along its nerve fibres move ten times faster than those in humans. Everybody knows how difficult it is to catch a cockroach.

When in danger some cockroaches roll themselves into a ball and this is usually sufficient to deter a predator. A cockroach found in Florida has developed an even more efficient weapon to survive. When under attack, it sprays its enemy with an irritating liquid.

Since cockroaches have a remarkable reproduction capacity it is very difficult to eradicate them completely, especially in tropical regions. Some cockroaches such as the German cockroach are even said to be 'born' twice. The female lays her eggs and then draws them back and keeps them inside her body until they hatch, when they emerge as minute offspring.

Although cockroaches are not well treated by people and everybody tries not to have them in their houses, some nations use dried cockroaches for healing purposes. Steeped in wine they have been used by the Peruvian Indians as a remedy to combat flu. The Negroes of Louisiana used cockroaches fried in oil to alleviate stomach troubles. Ashes of burned cockroaches are used in Jamaica in a form of ointment for skin diseases. Curiously, dried cockroaches are sold in some regions of Brazil as amulets to ward off evil spirits.

Although most people are familiar with black or brown cockroaches, there are many species which can be brightly coloured, but they are rarely seen because they live outdoors. While the house cockroach is relatively small, in some parts of the world there are some which are real giants. One particularly gigantic cockroach occurs in North Queensland and is called a giant burrowing cockroach (*Macropanesthia rhinoceros*). It is black and at first sight resembles a shiny black beetle. It can be up to 8 cm long and weigh some 30 g. It is rarely seen because it lives about 1 m underground and only appears on the surface to feed.

Strangely, Professor Harley Rose from Sydney University believes that this attractive cockroach is an excellent and undemanding pet. He sells them at a price of $65 a pair and believes them to be an excellent gift. But, as he recently admitted, he regretted that his cockroaches 'could not be taught to come when called.'

The Chinese Secret

A secret of international significance, which was kept undisclosed for almost 3,000 years is associated with a fascinating moth. The possession of this secret was of almost priceless value. The secret was kept by the Chinese, who discovered more than 2,500 BC that a caterpillar of a certain moth species found in China could be used for the production of precious silk, a material that was more valuable than gold. Such was the incredible story of the silkworm (*Bombyx mori*).

Although many caterpillars are known to produce a kind of silk to cover their cocoons, no other creature can secrete as long and as delicate a thread as the caterpillar of the famous silkworm. The thread is an unbroken strand which may be as much as 1 km long.

To reveal the source of the precious silk to strangers or to sell silkworm eggs to foreigners meant the death penalty. In order to add to the mystery of raw silk, the Chinese propagated various stories to confuse inquisitive foreigners. According to one, the precious silk threads were obtained by pouring water on to sheep fleeces under special climatic conditions which they claimed could only be found in China.

The silkworm was discovered by the Empress Lingchi, wife of the Chinese Emperor Huang-ti in 2640 BC. She was the first person to become interested in the extraordinary silky cocoons made by the caterpillar feeding on the leaves of the mulberry tree found in her garden. The Empress, the story says, ordered the silkworm eggs to be placed in special chambers and the hatched larvae to be fed with fresh mulberry leaves until they mysteriously covered themselves with silky threads. Her servants then reeled the silk threads and wove them into beautiful royal garments. The secret of silk making passed from one court to another and the Empress became so famous that she later became the Goddess of silkworms.

The Chinese guarded their secret very well and until the sixth century AD people in the west could only buy their beautiful cloth, because they had no idea of the source of this product. But in the middle of the sixth century, the Chinese monopoly was finally broken by two 'brave' monks. These monks had spent several years in China preaching Christianity and when they visited their homeland in Europe they not only brought back the secret of silk production, but also smuggled out a few live mulberry seeds. When the Emperor Justinian of Constantinople learned about the silk story, he persuaded the missionaries to make another trip to China to secretly bring back some silkworm eggs. Despite the threat of the dealt penalty hanging over them, the monks concealed some fresh silkworm eggs inside bamboo sticks, and were able to fool the border guards. In this manner they smuggled the precious eggs to Constantinople in AD 550. The brave European monks changed history by making silkworm breeding possible in Europe.

To ensure the best quality silk the Chinese elaborated special conditions for breeding the silkworm. They placed the eggs in special caves in the mountains where they found the humidity and temperature were most suitable. They achieved an incredible level of silkworm breeding and many treatises in the past dealt exclusively with silkworms. According to one, silkworms hate loud voices and for this reason they 'should have a peaceful, sheltered house, out of reach of unexpected south winds, and the damp vapours and exhalation of the earth . . . when the worms have just been born they are afraid too of the dust raised by sweeping.'

Strangely, in the old days in China women were not permitted to be involved in silk breeding because they were believed to be impure. They were not even allowed to approach the place where the eggs were collected. During certain phases of silkworm breeding, the men who worked with silkworms had to refrain from any association with women and husbands were not even allowed to talk to their wives. The silkworm breeders were only allowed to eat once a day, and their food could only consist of unboiled rice with some vegetables. The men also had to preserve their purity by abstaining from the use of a washerwoman and were not even allowed to visit a barber. The lower castes of people were not allowed to work with silkworms in an effort to keep the production undisturbed by their possible impurity. It was believed that if such rules were obeyed the caterpillars would develop well, but if one of these rules was broken, poor quality silk would be produced. At a certain stage of silkworm production the cocoons had to be submerged in boiling water to kill the tiny 'architects' so that they would not be able to emerge and hence destroy the precious silk. Until recently, every year in some regions of Japan memorial services were held for the souls of the dead silkworms, conducted by the local Buddhist or Shinto priest.

In this context it is worth mentioning two other amazing species of moth. The Indian moon moth (*Actias selene*) has the most sensitive sense of smell of all living creatures. It can detect its sexual partner from a distance of 5 km. Its antennae are so sensitive that they are said to be able to detect a single molecule of scent in the air.

Meanwhile, the newly hatched caterpillars of the polyphemus moth (*Antheraea polyphemus*) have a most incredible appetite. Only two days after they have started to feed themselves, they have been found to increase their weight over 80,000 times which is a real weight-gaining record.

Flying Fish

Without doubt, one of the unforgettable experiences of a sea voyage is to see a school of fish jump out of the water and 'fly' in the air like birds. Such a sight is most common on a calm night when these fish are streaming along swiftly beneath the sea surface like a galaxy of meteors.

These fishes, commonly called flying fish, do not actually fly because they do not possess wings which flap like those of birds. But they have expanded pectoral fins and in some species also pelvic fins which enable the fish to glide over the water surface. Before the fish starts its flight it swims extremely fast at speeds of up to 65 kmh, and then lifts itself suddenly into the air and glides for a few seconds, before diving back into the water. The flying fish usually skims just above the water surface, but in high seas, when the waves are huge, it may 'fly' some 3 m above the sea. Usually the fish covers a distance of about 50 m and the spectacle lasts only a few seconds. In favourable wind conditions a flying fish may glide for a distance of up to 400 m.

Sometimes the flying fish may be disturbed by the sudden appearance of a large predator from an unexpected direction and they may turn the 'wrong' way in their 'flight'. The school of fish may then hit a ship one after another with hammering taps. Sometimes a flying fish flies much higher than the rest of the school and, as a result, it lands on the deck of a ship. In his *Stories of Deep Sea Fish* Frank T. Bullen describes such an event. A man was once sitting on the deck and was suddenly hit by an unknown object and was 'howling like a spanked brat, and holding both hands to the side of his face from which the blood was trickling through his fingers . . . I examined his temple, finding indeed the marks of a heavy blow, blood and bruised flesh, and also two or three fish-scales.' It was later discovered that the aggressor was a 'splendid flying-fish about a pound and a half in weight, with its head almost driven into its body.'

As flying fish are attracted to light some fishermen, especially in Papua New Guinea, catch these fish at night using lanterns on their small canoes. Although this method is very efficient, it can attract other fish which behave in a similar way, by gliding over the water surface, and it can become very dangerous. For example, physicians were surprised when one day a man was brought to a local hospital with a strange fish embedded in his chest with its tail protruding like a stylet. The man soon died and was found to be wounded not by a flying fish but by another fish which also 'travels' at high speed above the water surface. It is called the needle fish and has a sharp pointed body and is able to catapult suddenly from the water reaching a considerable height in the air. It does not have expanding fins like those of a real flying fish but it can leap for some 5 m above the water. These fish are also attracted by fishermen's lanterns and although aiming at the light, they can miss and hit the man instead.

The best known is the Atlantic needle-fish (*Strongylura marina*). It has such a thin and sharp body that it looks like a sharpened pencil. It has an extraordinarily long snout which looks like the long bill of a bird. When such a fish catches its prey, usually a smaller fish, it first holds it crosswise between its needle-like jaws and then gradually transfers it into its mouth and swallows it whole.

There are certain fish in the Amazon regions which really can fly. They are called freshwater hatchet fish (e.g. *Hasteropelecus sternicula*) and have long pectoral fins and very deep chests housing lots of muscles which enable the pectoral fins to act as wings. As they fly, for up to 1.5 m, the flapping of the fins is so rapid that a buzzing sound can be heard.

The Vagabond of the Sea

Some fish have an extraordinary habit of using other large fish for transportation. Such are remoras (family Echeneidae) which attach themselves by a special sucking organ to large fish and travel great distances in this manner. Some remora species show a preference for large, deadly sharks as their vehicles and are called shark-suckers.

Remoras, which can be up to 90 cm in length, attach themselves not only to large fish but to any other large objects found in the water such as boats or ships.

The remora's peculiar habit was well known in ancient times. Pliny wrote that a ship's speed is strongly impeded if such a sucker fish attaches itself to the keel, and some remoras belong to the genus *Echeneis* which literally means 'holding back the ship'. The belief in the remora's extraordinary power to halt ships used to be so strong that they were even blamed for changing the course of battles. Caligula, the Roman emperor, blamed the loss of a battle on remoras 'holding back' his ship. The effort of 400 seamen to free the ship from the fish's powerful grip was unsuccessful. This story cannot be true, but the power of remoras to delay the movement of vessels was considered to be true until the fifteenth centrury in Europe. The remora's sucking power has been utilised by various nations since ancient times. Christopher Columbus was surprised when he saw a peculiar method of catching turtles in America. He reported that the Red Indians 'would tie a line to its tail, and throw the fish into the water where they knew turtles abounded. The remora would attach itself to the first turtle encountered, the fish and its "catch" then being hauled ashore.'

In many regions remoras are still used for catching fish. For this peculiar 'angling' method, the fisherman attaches a long line to the remora's tail and allows the fish to swim off. When it finds a large fish it attaches itself to it and the line is then hauled back. In this simple way the 'hunting' remora and its catch are both brought straight into the fisherman's hands. Only large remoras are used for this purpose. Smaller remoras, however, are also occasionally used in the West Indies, not by professional fishermen, but by small boys. Their method is simple. They look for fish from a boat and after spotting one they put their remora into the water so that it attaches itself to the fish and catches it for them.

Fishermen who keep large remoras for fishing purposes develop cordial relations with their 'hunters' and they become real pets. They are treated with respect and given the best food. The owner even talks to his 'pets' especially before hunting, when he begs the fish to bring him a good catch. Some fishermen believe that remoras are very intelligent and that they understand and follow human commands.

Miraculous power is attributed to remoras in some regions. A potion prepared from these fish was believed to stop aging in women. It also used to be believed that remoras had the power to slow down legal proceedings and 'slow down the course of love' and were also believed to have a remarkable ability 'to draw up gold which has fallen into a well, however deep it may happen to be'.

Perhaps most peculiar is the use made of remoras in some regions of Madagascar. When a woman is found to be unfaithful to her husband, remoras are used as a strange form of punishment. Dried pieces of the remora sucking organ are placed around the woman's neck in the belief that, like a remora, she will now firmly attach herself to her husband.

The remora hates being alone so much that it will attach itself to any object, including the body of a swimmer. Skin divers are occasionally frightened by these fish which try to attach themselves for a free ride. The urge for company is so strong that when many remoras meet they immediately attach themselves to each other. With the largest on top and smallest at the bottom they form a peculiar pile of 'plates'. Such a 'multifish' object swims slowly and in a rather unusual manner.

The habit of attaching itself to the body of a big fish is very advantageous to remoras. It provides them not only with a free ride, but also with protection as no sea predator is likely to attack such a strong host. The attached remora also has a chance to feed from the same table as it can occasionally catch scraps of food left by its host.

Several remoras may attach themselves to the same big fish. A fisherman was once astonished to find a remora attached and still alive in the mouth of a large shark he had killed. Remoras must perform a cleaning duty or their presence in good shape inside a big fish-eater would be difficult to explain.

The disc-shaped sucking organ of the remora is a specially modified dorsal fin on top of its flattened head. This disc is formed of a series of transverse, tooth-studded plates. The remora attaches itself by creating a strong vacuum between the disc and the body of its chosen partner. The sucking power is so strong that it is impossible to dislodge the fish by pulling and it can only be detached by sliding it forwards. But the remora can detach itself at will and leave its 'vehicle' to hunt alone for some time.

Jesus Cristo Lizard

Among the curious lizards which inhabit Central America are the basilisks (genus *Basiliscus*). The male of these lizards, which can be some 80 cm long, is very strange. It usually has erectile crests on its head and on its back. When the animal is excited, it raises its crests and the skin under its lower jaw greatly expands so the animal looks fierce enough to frighten its predator. This remarkable display was the reason it became a symbol of ferocity to local people.

It was even believed that such an animal could kill with its horrible breath and it was named basilisk after the legendary fire-breathing monster. In fact these animals do not deserve such a reputation as they are entirely harmless to men.

In the past, however, the basilisk was a legendary animal regarded as the king of serpents. It was depicted with a serpent's tail and people were afraid even to look at it. Basilisks were thought to be the offspring of a cockerel and a snake.

The basilisk can run across water as if on the ground. When frightened it runs on its hind legs and escapes from its enemy by racing for water where it 'walks' on its surface, making distinct sounds and for this reason the animal is called 'tetetereche' by local people.

It seems a miracle that such a reasonably large animal can run across water without sinking and this is why it is known in Central America as the Jesus Cristo, refering to Christ's miraculous ability to 'walk on water'.

It has a fringe of scales on its long and wide-spread hind toes which enable the lizard to skim across the water. It is its speed, however, of up to 12 kmh, that enables this animal to perform the 'miracle', without sinking. It is said that one basilisk was seen to cross a lake some 400 m wide.

The sight of a basilisk running across the water on its two hind legs is really amusing. As it slows down the basilisk finally begins to sink and becomes submerged, swimming normally for some time.

The largest and probably most spectacular among basilisks is the green crested basilisk (*Basiliscus plumbifrons*) found in Costa Rica. The male of this species has a very large crest on its head, another along its back and even a crest on its tail.

Some basilisks change colour with great ease. The banded basilisk (*Basiliscus vittatus*) is dark brown during the day, but at night when resting its lateral bands become vivid yellow or even white.

Fishy Anglers

There are some strange fishes that have a peculiar habit of fishing for other fish using bait like human anglers. Some angler fish even have a fishing rod, which evolved from the first ray of their dorsal fin. In some species the fishing rod can be four times the length of the whole fish. At the end of the line there is a bait resembling a worm which even temptingly vibrates. The fish's clever angling strategy is different from that used by a fisherman. Its 'worm' is not designed to be swallowed as there is no hook attached to the lure. When a small fish approaches, the 'fishing' fish draws back its line slightly. By the gradual withdrawal of its bait the fish plays with its prey until it is drawn close enough to its mouth to be snapped up.

In very deep seas a bait described above would not be sufficient to attract fish because of the lack of light and so the deep-sea anglers have 'invented' luminescent bait. Some, such as those of the genus *Linophryne*, are not only equipped with a shining bait, but also have a peculiar tree-like outgrowth hanging from their chins which is also luminescent. Although the luminous bait is usually on the outside, in the deep-sea fish belonging to the genus *Galatheathauma* the luminous bait is inside the mouth. The fish hunts with its mouth open ready to catch any inquisitive small fish. Some deep-sea angler fish, which live on the sea floor at a depth of some 3,500 metres, have such distensible stomachs that they can swallow fish three times their own size.

One of the largest angler fish, the Lophiid angler (*Lophius piscatorius*), which can reach 2 m in length, is found in both the North and South Atlantic. This fish is a peculiar shape, like a huge disc covered with curious outgrowth. It is so ugly that some fishermen call it the sea devil and, because of its enormous mouth, it is also called the fishing frog. Although these fish normally catch fish lying on the sea bottom, they occasionally come to the surface where they are even known to catch and devour seagulls and other sea birds. For this reason they have gained another popular name, 'goose fish'.

The Lophiid angler does not have a fishing rod and its bait is on its head in front of its eyes. By gently waving its lure, the fish attracts passing fish. When the curious ones approach the lure within striking distance, the angler fish's mouth suddenly opens and the inrush of water sweeps the prey into its cavern-like throat. This fish is so voracious that on a trawler it gorges itself on its fellow captured fish so it usually dies with a full belly after all.

Perhaps the most unusual way of catching fish is employed by the so called decoy fish *(Iracundus signifer)*. It does not possess a rod nor a worm-like bait and it lures the fish in another, very peculiar way. Its dorsal fin has evolved in such a way that it can be changed to look like a small fish. The hunting fish lies at the sea bottom hidden in the sand except for its fish-like fin, which is exposed to attract other fish. Furthermore, the fish-like fin becomes red in colour and even has an eye-like spot on its 'head'. This 'fish' attracts larger fish which come for any easy catch, but when they try to swallow it the hidden owner snaps up its prey. After making its catch the fish alters its dorsal fin to its normal shape and the red colour fades.

Aboriginal Totem

Dolphins are the best known of all the toothed whales. Some scientists believe that dolphins may have a potential IQ equal to that of humans and that they have not developed a civilisation similar to ours because their perfect adaptability to their environment makes such additional development simply unnecessary. These scientists base their belief on the dimensions and characteristics of the dolphin's brain. The brain cell count of a dolphin per cubic centimetre is the same as that of humans. The study of the brain of the bottlenose dolphin (*Tursiops truncatus*) has revealed that it enables the animal to live without sleeping by putting each of the two hemispheres of the brain to 'sleep' independently. As a result when one half of the brain is asleep the other is keeping watch.

Dolphins can be trained to perform many functions because of their high intelligence. They have even been trained to carry explosives to blow up enemy submarines. According to Michael Greenwood, a former US Navy and CIA scientist, dolphins were used to kill Vietnamese frogmen during the Vietnam War by striking them with syringes 'loaded with a high pressure gas that caused the lungs and stomach cavities to explode.' The US Navy admits that they trained dolphins to be used as divers' guides to carry explosives and mines or detect submarines but denies having used dolphins in actual warfare.

Dolphins are also fascinating because they demonstrate a high degree of altruistic behaviour, not only towards their own mates, but also towards other animals and humans. For instance, a 1949 report documents a Florida woman being saved by a dolphin. She was alone and drowning when suddenly a dolphin arrived and carried her to the beach, evidently saving her life. There are many stories of close friendships established between these intelligent creatures and humans. One young female dolphin became so friendly with a 12-year-old girl that they always played together in the sea. She used to put small children on the dolphin's back to ride as if on a horse. This dolphin, named Opo, became famous and thousands of people went to the small township of Opononi in New Zealand to see the famous Opo. She learned to play ball and could hit the ball high into the air with her tail. Tragically, this famous dolphin was accidently killed by a gelignite explosion. To commemorate Opo, a monument representing the famous dolphin was erected in the town.

Anther dolphin, named Pelorus Jack, used to escort ships passing through Pelorus Sound, which is a very hazardous stretch of water between the North Island and Nelson in the South Island of New Zealand. The captains of the vessels were very pleased with this unusual assistance and thanks to their excellent navigator, there had been no accidents since 1888. Then, in 1912, the famous pilot dolphin suddenly disappeared. It is speculated that, despite the animal being protected by law, he was probably harpooned by a whaling ship.

Dolphins can locate obstacles in the water with amazing accuracy, even with their eyes closed. With half a tennis ball placed over each eye the animal behaves as if it still could see perfectly. We now know that dolphins use an ultrasonic detector to receive and send echoes bounced from obstacles in the water. While a man cannot distinguish between a wooden or metal ship from a distance, a dolphin can differentiate between two different fish of similar size and appearance.

Scientists have always been puzzled by the variety of sounds made by dolphins when communicating with each other. They have studied their clicks, grunts and whistles with special microphones and dolphin 'conversations' have been taped. Some biologists believe that in the future we will be able to learn the animals' language, which will help us better understand these unique animals.

Dolphins have figured prominently in the beliefs of various nations because of their unique behaviour. In ancient Greece, a dolphin became a talisman for the dead as dolphins were believed to assist people travelling to the after-life. For this reason a dead person was buried with a dolphin token placed in his mouth. Dolphins also became a symbol of diligence and love and dolphin charms were used by travellers on sea and land.

River dolphins were also highly respected in South America. The Amazon dolphin, also called boutu (*Inia geoffrensis*), is regarded by local Indians as sacred and cannot be killed. There is a peculiar belief among these Indians that the dolphin is very fond of young girls and during carnivals disguises himself as a man. If an unmarried girl becomes pregnant these Indians believe it is to be the result of a dolphin's love and the girl is not disgraced in the community.

Dolphins are also held in great respect among the Australian Aborigines. Some tribes, living in the islands near Arnhem Land, hold the bottlenose dolphin as their totem and do not harm them. They have excellent relations with these dolphins and the young Aborigines learn from their fathers how to communicate with them in order to obtain their help, especially when fishing. As one Aborigine reported: 'I used to go out fishing with my uncle who had studied the dolphins for many years. When he called them usually three dolphins would appear at the sides of the boat and one at the back. They would then make a field of sound and drive a whole mob of fish up onto the shore.'

There is a common belief in Vietnam that dolphins were created by the god of waters and have the power to protect seamen and rescue them when they ask for help. No wonder that a stranded, dead dolphin is always buried with great respect. It used to be customary for the person who found such an animal to wear a mourning turban. It had to be worn for three months and then burned. After mixing its ash with the bones of the dolphin it was respectfully placed in a sanctuary.

16.
Unbelievable Defence

Pharaoh Emblem

Although the Egyptian cobra *(Naja haje)* is less spectacular than the hooded cobra found in India, it was the most venerated snake of ancient Egypt and was the national goddess of Lower Egypt in ancient times. A cobra emblem was found in the famous tomb of the Pharaoh Tutankhamen, excavated in 1922, and this indicates that this snake was also an imperial symbol.

The snake goddess was frequently represented in the form of a cobra spitting fire and the cobra image was placed on the pharaoh's forehead in the belief that it would protect him from his enemies. The Egyptians believed that the four corners of paradise were guarded by fire-breathing cobras.

These snakes were believed, like humans, to live after death and for this reason the snakes were even embalmed before their funeral. The cobra was also regarded as a symbol of fertility in ancient Egypt and amulets in the snake's form were worn.

These snakes were even occasionally given to some prisoners since being bitten by the deadly snake was considered an honourable way to die.

This snake also plays an important role in the beliefs of certain tribes in other parts of Africa. On Fernando Po, an island off the coast of West Africa, a special festival is organised in honour of a particular cobra which is worshipped and regarded as the guardian of the whole community. It is believed that the cobra-god can bring various kinds of blessings. During an annual ceremony a cobra skin is hung in a public place in each village. Mothers bring their newborn babies to touch the tail of the sacred snake in the belief that it will bring them luck.

Cobras are now a popular tourist attraction in Egypt, and appear in staged fights with mongooses. Although these animals do not attack each other in the wild, when placed together in a cage they fight fiercely. The mongoose always wins, but before the contest is over and while the animals are still alive the show manager stops the fight because if the precious 'artist' is killed and the other animal seriously wounded, it would be difficult to find replacements for the next show.

Perhaps the most interesting cobras are the spitting cobras, such as ringhals (*Haemachatus haemachatus*). People are more wary of these attractive snakes than other cobras because they are dangerous even without biting. When this cobra is disturbed it uses an extraordinary weapon. It jets liquid poison straight into the enemy's eyes. The snake shoots with remarkable precision from a distance of up to 3 m and rarely misses. It seems to be aware that the poison is only effective in the eyes. The poison causes excruciating pain and irritation which is so serious that if the venom is not flushed immediately, it can even cause blindness. Both people and animals can be affected if they walk past and disturb a spitting cobra.

Spitting cobras are widespread throughout Africa and some people have been permanently blinded as a result of this snake's method of defence. They like to live near humans and are often found to live even in big towns. When we lived in Nairobi in Kenya several spitting cobra attacks on humans were reported in the city. Special warnings were given to the public to avoid walking too close to hedges and provoking a hidden cobra to attack.

Biologists are often puzzled why this snake wastes its poison when it would be more efficient if it used it like other venomous snakes. This snake can, of course, bite and its poison is among the most potent. It has been estimated that one gram of the dried venom could kill 160 persons. The snake only bites, however, to kill prey for food. For defence it uses its shooting technique. Unlike other snakes, the opening of its venom glands is at the front of the fangs so the snake can direct its poison stream forward after taking a proper aim at the target. To eject its venom the snake applies muscular pressure to the venom glands. The snake may spit several times in rapid succession if he misses his target the first time.

Those who collect the poison from spitting cobras must wear diver's goggles to make sure that their eyes are well protected.

Small Peugeot

When William Bebe, the American deep-sea explorer, descended into the lower depths in his bathysphere he was amazed at the enormous number of illuminated sea creatures, and reported that their lights 'were bewildering." He was particularly impressed with one fish, which he named the 'aristocrat of the deep sea'. He wrote: 'Along the sides of the body were five beautiful lines of light . . . each line was composed of a series of large, pale yellow lights . . . In my memory it will live throughout the rest of my life as one of the loveliest things I have ever seen.'

Each species of deep-sea fish distinguishes its friend from its enemy by the light emitted and the colour pattern, which is not an easy task in the complete darkness of the deep sea. Some of these fish, called lantern fish, use their light not only as a method of communication but also as a defence weapon. When a big predator attacks, the lantern fish emits an intense light from its luminescent organ on the front of its head so that the aggressor is dazzled and looks for another victim.

It has been observed that many deep-sea fish have a distinctive blinking pattern which they seem to use as a means of communication. It would also appear that males flash their lights to lure predators away from their female partners and in this unique way ensure their survival. In some species only the males send flashing signals while the females of the same species do not flash in response, but are able to find their partner easily.

Some fish species blink at a rate of two to three flashes a minute, but when under attack increase their flashing rate dramatically. They may flash as many as 75 times a minute and this effectively confuses the predator. The luminescent organs are located in all sorts of different places. In some species they are not only outside but inside the body. The lights of a deep-sea fish of the *Chauliodus* species are inside its mouth. Over 300 light dots illuminate its whole mouth so perfectly that it uses its lighted open mouth as a lure. Any curious fish which enters the lethal opening is immediately lost in the 'cavern'.

The most spectacular luminous fish is only some 8 cm long, of the genus *Photoblepharon* and occurs in the Red Sea. Its luminescent organs are under each eye and they emit probably the brightest light of any living creature. In the darkness of the night the two lights resemble car headlights and French divers nicknamed this strange fish 'le petit Peugeot'. In addition, the fish has 'eyelids' which it can raise to mask its light organ very efficiently if it wishes to become less visible to predators.

The light organs, known as photophores, have a very interesting structure. Each organ is literally packed with a special kind of bacteria which can emit light. These bacteria live in a perfect symbiosis with their fish host. They are nourished by the fish's blood supply, which carries sugar and oxygen, which enables the bacteria to give off light. As the blood vessels supplying the light organ can be constricted by tiny muscles, the fish can shut off the blood supply and so extinguish the light at will. To strengthen the emitted light it is passed through a special layer of silvery crystals which concentrate the light like a telescope's mirror concentrates starlight. In addition, the inner walls of the light organ are lined with a tissue containing black pigment, thus preventing the fish from being blinded by its own light. A newly hatched fish of this species has no luminescent bacteria in its light organ so it must find some tiny partners and 'offer' its partnership. Fortunately, there are abundant luminescent bacteria in the sea also looking for suitable partners. The partners live together until the fish dies.

The most strange looking fish among the luminous fish are no doubt the marine hatchetfish (*Argyropelecus* species). Their light organs are located at the lower end of the body. These tiny fish (most species just 3 to 5 cm long) resemble a hatchet, hence the name of the fish. With its huge bulbous eyes and large gaping mouth it looks like a suffering distorted human skull.

Some fish, however, do not 'employ' bacteria. Their luminous organs secrete a phosphorous substance called luciferin which glows when broken down by the action of oxygen in the presence of an enzyme. The oxygen is supplied by small blood vessels surrounding the gland. By simply constricting the blood vessels, the oxygen supply can be cut off and the light 'switched off' at will.

The luminescent organs of many fish continue to emit light for several hours after they have been cut off the fish's body. Natives of the Banda Islands use light organs as bait and claim to have good catches with this rather unusual bait.

Bicycle Lizard

One of the most spectacular Australian lizards is the frilled lizard (*Chlamydosaurus kingii*). This fairly small lizard has evolved a peculiar defence technique. Unlike other lizards, it has a large frill, which normally hangs in loose folds around its neck, but when it is cornered it efficiently uses its 'scare organ' to frighten the attacker. It quickly opens its mouth and as a result the frill raises up like an umbrella. By opening its mouth wide, the lizard shows its vivid yellow lining and hisses as if it would attack its predator. This spectacular 'umbrella' display has its effect as the predator is frightened by the suddenly 'enlarged' lizard and usually withdraws, allowing the lizard to escape unhurt.

This unique defence method is so effective that even dogs, which usually attack and often kill much bigger lizards, are scared by the frilled animals and leave them alone. Allan Cunningham, who was one of the first scientists to study the behaviour of this species, called it 'a lizard of extraordinary appearance' and said that 'there is indeed nothing quite like our frillneck anywhere else in the world'. There was a great sensation when this lizard was exhibited for the first time at a scientific conference in London in 1893. One zoologist was so amazed by its bizarre appearance that he followed the creature 'on hands and knees to watch it carrying round on its hind legs and displaying its frill'.

Not only the lizard's peculiar frill display is extraordinary, but its gait is also most unusual. It runs fast on its two strong, hind legs in a strange, erect position with its front legs raised in the air and with its large head moving from side to side. This extraordinary gait has given the animal another popular name 'bicycle lizard'.

This peculiar lizard, which occurs in northern Australia, is well known to Australian Aborigines. They have a myth that this lizard was once responsible for causing a great flood which destroyed nearly all living creatures in the region, because it sang a secret prayer song to the Great Spirit responsible for making rain.

The frilled lizard is a typical forest-dweller and climbs trees with ease, searching for insects which are its main food. The lizard, which can be up to 90 cm long, has a long tail which may account for about two thirds of its total length. Large lizards of this species are even known to attack birds and devour them. For instance, they have been known to steal hens from farms near forests.

This lizard is a very good hunter and it is known to be an effective ambusher. Posting itself quietly on a tree trunk, the frilled lizard patiently waits for a small animal to pass and then jumps down to the ground and catches its prey.

These remarkable animals used to be very common, but their number has diminished dramatically in recent decades. This is not due to predators, but to the destruction of their habitat. With the destruction of the tropical forests, the frilled lizard, which is a tree dweller, loses its only home.

Frilled lizards used to be exported as pets, especially to the USA. Before this was banned, as much as $100 was paid for one live frilled lizard. These strange lizards have become so popular in Japan recently that they actually seem to be more popular than the famous koala. It often appears on Japanese television and toys with the characteristic expanding frill have become popular. As one author noted: 'This little Aussie intrigues foreigners, in particular the Japanese, with whom it has become a craze. With the advent of wind-up toys in the likeness of the frilled lizard, the stuffed koala has been consigned to the cupboard.'

Thorny Devil

One Australian lizard has an impressive scientific name — *Moloch horridus*. Moloch was a mythical Canaanite god who demanded child sacrifice, and 'horridus' literally means 'horrible'. One would therefore expect an animal with such a description to be the most terrible beast, but despite its unique, frightening appearance, this creature is completely harmless to man. Although it is small, only some 15 cm length, its body looks most formidable. Its dorsal surface, its head, tail and legs are covered in sharp spikes and hence some biologists call it the porcupine lizard. Its general appearance is so frightening that the lizard has been called the thorny devil or moloch.

It is among the most sluggish of all lizards. It may take an hour to move only a few metres and is motionless most of the time. As such a slow-moving animal is unlikely to escape from a predator, its only chance is to frighten its enemy with its thorny armour. When under attack, this weird lizard hides its head between its front legs and raises a thorny hump on the back of its neck instead of its head. This is a simple but effective technique, since no predator will attempt to bite a mass of sharp thorns.

Its remarkable armour is only developed for protection as this creature does not attack other animals, except for certain ants which are the animal's main food. Aborigines call this lizard 'mingari' from the word 'monga' meaning ant. The clever lizard only eats those ants which do not possess painfully nipping jaws. It has been found that the moloch feeds exclusively on three species of ant of the genus *Iridomyrmex*. Early in the morning it positions itself in the path of the marching ants and catches them with its sticky tongue. The moloch is an elegant eater as it leisurely picks up the ants one by one. It usually eats about 40 ants a minute, but is a big eater and may devour as many as 2,500 ants in one sitting. After such a big meal the lizard may not eat for several weeks.

The thorny devil is wonderfully well adapted for living in desert conditions. This animal can literally drink water from its own body. During cold nights in the desert, dew appears on the lizard's skin which is covered with thousands of tiny grooves. The water is not absorbed through the skin, but gradually runs to the animal's lips and is swallowed.

The thorny devil spends its nights hidden in burrows or under rocks. Its eyes are unsuitable for nocturnal excursions. During the day these creatures do not only rely on their formidable appearance to deter predators, but also try to remain unnoticed. The animal can change its colour perfectly to match the background, although not as quickly as a chameleon. Depending upon the surroundings, this strange animal may even develop black markings on its body to resemble the terrain.

Australia is also the home of another weird-looking lizard called the bearded dragon (*Amphibolurus barbatus*), which is famous for its spectacular display. This lizard is some 60 cm long. When endangered, it also uses a trick to frighten its enemy. It inflates its spiky beard and rib section by opening its mouth wide and revealing the vivid yellow lining. This, together with the beard, looks formidable, and such a display of power is usually sufficient to deter the predator. If its display fails, however, this lizard bites its opponent. The spectacular display is also used by the males to attract their sexual partners during courtship.

Another interesting lizard is the Texas horned lizard (*Phrynosoma cornutum*), found in North America. It has peculiar horns on top of its head, as well as numerous spiny projections on its back. Its body is short and flattened and looks like a toad, hence another name of this grotesque lizard, the horned toad.

This strange-looking lizard has an interesting ability. It is known to eject blood from its eyes when molested and when captured it is said to become so unhappy that it starts to 'cry blood'. The first Spanish explorers who saw this unusual lizard 'crying' were astonished by this behaviour. It does not only secrete blood when molested, but is able to squirt it in a fine jet to a distance of up to 1 m.

Some species of horned lizards can change their shapes remarkably when under attack. They inflate their bodies and hiss frighteningly until the attacker usually leaves them alone.

It is not easy to spot these lizards in the wild, stony ground where they live, because they match their surroundings almost perfectly. When such a lizard closes its eye lids it virtually disappears and the camouflage is so good that it is only visible when it starts to move.

Grave Digger

There is a curious mammal which has the armour scales of an alligator, the eyes of a pig and the ears of a mule. This creature is called the armadillo, and all species, except one which lives in the southern parts of the USA are found in South America. When the first Spanish explorers saw this striking animal, which is almost completely enveloped in a hard shield of armour, they called it armadillo meaning 'little armoured thing'. With tough, horny plates covering its back, sides, front of the head and occasionally even its tail, the animal looks like a prototype of a modern war tank.

The now-extinct relatives of the present armadillo had enormous body shields measuring up to 3 m. The American Indians used these shields as shelters or roofs for their huts.

The Maya Indians had an interesting myth about the origin of these strange-looking animals. According to this myth, when a vulture became very old it changed itself into an armadillo. When it was ready for the miraculous change, it went into a hole in the ground where it began its transformation. The bird's wings completely disappeared and the creature lost all its feathers and eventually turned into an armadillo with its armoured body. The Indians' belief is understandable as there is some similarity between the bald head of the bird and that of the armadillo.

With its strong claws and remarkable armour, this animal seems to be created to fight, but in fact it is a timid animal which usually does not show any aggressive behaviour. Their armour certainly does not attract many predators, but, in addition, one species of armadillo can roll itself into a ball.

The Argentinian three-banded armadillo (*Tolypeutes triticinctus*) can make a perfect and impermeable ball, closing its shell like a steel trap, when attacked, which is sufficient to discourage any predator.

Other armadillos use another effective method of defence. When approached by a predator, they seek safety by burying themselves in the ground. If the soil is soft enough, the animal disappears under the earth with such remarkable speed that the attacker has no chance of catching it. Armadillos are known for their 'passion' for digging and for this reason in the Mississippi region these strange animals are called 'grave diggers'.

Armadillos are also known to seek refuge in water when attacked and, if they can reach the water in time, have a good chance of survival as they are excellent swimmers. It is rather surprising that an animal with such heavy armour is able to swim. The Indians used to say that the heavy armadillo walks along the bottom of the river as it cannot swim. In fact, the armadillo is well adapted to such an emergency. He simply gulps air, which inflates its intestines, until it is sufficiently buoyant. As the animal can also stop breathing for almost six minutes it is a good diver.

Armadillos feed mostly on insects, especially ants, but they also catch other animals. In the stomach of one armadillo a toad, ten tarantula spiders, a small snake and a scorpion were found. Some armadillos use their strong claws to break open termite mounds to eat them. The best in this respect is the giant armadillo (*Priodontes giganteus*), which can measure up to 1.5 m and weigh up to 60 kg and lives in tropical forests of South America. It has powerful claws on its forefeet with which it strikes the hard termite mound and collects the termites with its sticky tongue. Whole termite hills are occasionally found destroyed by this animal.

The smallest and most peculiar is the pigmy armadillo of Argentina, also known as pichiciago (*Chlamyphorus truncatus*). It is only some 13 cm in length and lives in burrows like a mole. This animal has pink armour, which does not completely cover its body and is only attached along its back, and its rear end is protected by a kind of separate shell or plate of armour. As this burrowing armadillo is most likely to be attacked from the rear when escaping from a predator, this rear plate serves as a shield to plug its burrow and prevent further attack. This is an excellent protective adaptation, well suited to the habits of this armadillo.

South American Indians like armadillo meat, regarding it as better than pork, and hunt armadillos, usually using trained dogs. As the dogs may kill the animal, the hunters usually intervene as they prefer bringing the prey home live. During the time of the Aztecs in Mexico, live armadillo were sold as food or exchanged for precious cocoa beans.

Roasting an armadillo is rather a strange procedure as the whole animal, including its armour, is cooked in the fire. When the meat is ready, the shell provides a ready-made plate. An empty shell can be used as a basket or for making a musical instrument — when fitted with a neck and with strings attached it becomes a guitar.

Their extraordinary appearance has made armadillos very popular animals, especially among children in South America and in parts of the USA. Many comic books and T-shirts picture the famous armoured animal. Even jewellery is made in the form of an armadillo. Special events known as Armado Festivals are held annually in some regions of South America.

Armadillos have even become pets, although one handicap has prevented them becoming popular. They emit an unpleasant, musky odour from their scent glands near the base of the tail.

The nine-banded armadillo (*Dasypus novemcinctus*) was found recently to have one rather unexpected quality which made it important in medical circles. This strange animal is similar to humans in that it can contract leprosy, a disease for which there is still no vaccine. Since these animals can live for up to 15 years in captivity, they are almost ideal animals for the study of the disease. Scientists believe that research with this unusual animal may now help develop a vaccine against this deadly disease.

Eccentric Cucumber

Despite its name, the sea cucumber (class Holothuroidea) is actually an animal related to the starfish. They usually have a long cylindrical body and leather-like skin, which can be likened to a cucumber, hence its popular name. These very primitive echoniderms have a distinct mouth, surrounded by a ring of tentacles. At the other end there is an anus which, in some species, is used for breathing. These creatures suck water through the anus which then goes into their huge respiratory tree.

There are about 900 sea cucumber species which vary in size and colour, and can reach a length of up to 1.5 m. These animals are widely distributed in almost all seas, often living at great depths. Most sea cucumbers are floor-dwellers, often lying in the sand, while others are modified for swimming or floating.

Some sea cucumbers have developed a very strange method of defence which puzzled biologists. When the animal is under attack, it shoots parts of its own body at its predator. The animal splits its body wall and ejects gut and respiratory trees at the predator. While the attacker is busy with this easy meal, the clever animal escapes. Amazingly, its missing internal organs regrow later and the animal can again use its rather 'painful' trick.

Even more remarkable is the defence adaptation of some sea cucumbers, which prefer not to offer their entrails in exchange for life. They have developed special structures called Cuvierian organs, sorts of tubules branching from the bases of the animal's respiratory trees. When under threat, it bends the hind part of its body, takes aim and shoots tubules through its anus. The sticky tubules become entangled with the predator and before it can free itself the animal escapes unharmed.

Some sea cucumbers apply an even more unusual trick to fool predators. The creature suddenly breaks in two and, before the confused predator can decide what to do next, the two 'miraculously' born animals escape. The lost parts of the body of each animal eventually regenerate.

Other sea cucumbers are even known to possess chemical weapons which they use mainly for hunting. They secrete a poison which paralyses and kills small fish. People living on some Pacific Islands are aware of this and exploit the sea cucumbers in a peculiar method of fishing. They cut the sea cucumbers into small pieces and throw them into the water. Suffocating fish soon float to the surface where they can be easily collected. Since the poisoned fish are edible and do not harm humans the method is very useful.

Sea cucumbers can become the homes of fish. A small, elongated fish called the pearlfish (*Carapus apus*) enters the sea cucumber by inserting its tail first so that only its head protrudes from the host's anus. What is more surprising is that the strange guest is not only happy with the shelter but also eats some of its host's internal tissues. This does not threaten the sea cucumber's life however. The fish only feeds on its host for a short period of time. Once it matures it ignores the host's flesh and seeks food outside. The fish only leaves its shelter at night and remains inside during the day for security.

Despite the fact that sea cucumbers are rather repulsive in appearance and taste rather unpleasant, some species are considered a delicacy. The Chinese were the first to discover that this strange animal can be eaten and they use it for making a delicious soup. The sea cucumbers are smoked and cured and after they dry they become as hard as wood. By boiling these 'wooden' pieces, popularly known as cucumber 'meat', a gelatinous, tasty soup is made. The Chinese believe that sea cucumbers are not only a delicious food but are also a sexual stimulant.

In the Pacific islands sea cucumber, often called bêche de Mer, became an important source of income for the native population. Demand for the lucrative, Chinese market was so strong that early European traders in Australia forced islanders to process the precious commodity for export. The islanders were virtual slaves, forced to work hard gathering and processing sea cucumbers. These people were so mercilessly exploited that in 1805 the Governor of NSW issued a special order prohibiting 'inviting' the islanders to Australia.

The Australian Aborigines were also affected by the sea cucumber industry. They were forced to work in specially organised camps called curing stations. They were so unhappy with their conditions that, on occasion, they destroyed the curing stations and killed the traders. They occasionally mutinied on the ships carrying the final product, throwing the hated sea cucumber traders into the sea. In this way an Oriental delicacy led to the incredible misery of many people.

Barking Lizard

Some people are frightened at night by a sudden, mysterious call, 'gecko, gecko, gecko', in a voice closely resembling that of a man. It is hard to persuade them that the calls are made by a species of small lizard, known as geckos, namely the common house gecko of North Africa (*Hemidactylus mabouia*). Many other geckos bark or chirp and hence these lizards are often called 'barking lizards'.

Some geckos have such loud voices that they can be heard at a distance of up to 100 m. The African whistling gecko (*Ptenopus garrulus*) are famous for the deafening noise they make when they call simultaneously. The kind of sound uttered varies with the species — one emits a sound like a crying baby. In some Asian countries there is a widespread belief that if a gecko calls from the bedroom of a newlywed couple, they will have a long and happy life. Strangely, a few species of gecko are mute.

Despite being harmless creatures feeding mainly on insects, Australian Aborigines fear them profoundly and they gave them the frightening name of 'wood adders'. The reason for this strange attitude is probably because some species (such as *Diplodactylus intermedius*) squirt a viscous substance from the spines of the tail, which may cause painful irritation if it gets into the eyes.

Most geckos are very swift lizards so they have a high chance of escaping from most of their enemies. But, when unsuccessful, the gecko has a special trick to save its life, offering the predator its tail. These lizards are specially adapted for this trick as their tails break easily. The disembodied tail wriggles convulsively on the ground as if alive and this attracts the enemy's attention, enabling the gecko to run away. The tail regrows but is usually shorter and paler. The new tail cannot be discarded, as it lacks vertebrae and is made of cartilage tissue. Shedding tails must be very common, since it is claimed that in some areas only one in ten geckos have normal tails. Sometimes the original tail does not break off completely and a new tail may grow on one side, resulting in a 'two-tailed' gecko. The Australian marbled gecko is known occasionally to have three tails and such a triple-tailed gecko is an extraordinary sight.

Some geckos' tails are so fat that they resemble caterpillars. When the enemy becomes interested in this 'rich' meal it grasps the 'caterpillar', losing interest in the lizard which escapes, shedding its tail. Some fat-tailed geckos are rather reluctant to offer their tails for life. When under attack a fat-tailed gecko (*Diplodactylus conspicillatus*) found in Australia escapes to the nearest burrow and then inflates its tail with air, closing the entrance and forming an impenetrable barrier.

The flying geckos of southeast Asia can avoid capture by gliding to a safe place. They have extensions of skin along their bodies and tails which makes such a 'flight' possible.

Most geckos are remarkable climbers and some are real acrobats. They can walk upside down on a smooth ceiling and not drop off. They can hang upside down from a sheet of smooth glass — an ability that fascinates biologists. We know that the secret lies in their special toe structure. On the underside of each toe there are pads consisting of hook-like bristles which enable the animal to grip firmly, even to a smooth glass surface. It has been revealed that, because of these microscopic clinging bristles, the gecko's toe pad grips firmly, even on a smooth glass surface, because the glass has sufficient minute irregularities for the bristles to attach to them. The only exceptions are very highly polished surfaces.

17.
Astounding Military Strategies

Incredible Tongue

Among the bizarre lizards of the Old World, the most peculiar is the chameleon which looks like a miniature version of the ancient dinosaurs. One of the most interesting is the Jackson's chameleon (*Chamaeleo jacksonii*) which is common in some parts of East Africa. When we lived in Nairobi in Kenya, we once spotted a beautiful specimen in our garden. To observe this interesting, three-horned lizard we brought it home. When our African servant saw our 'pet', however, he was as terrified as if he had seen a devil. He ran away and it was difficult to persuade him to come back to the house. He only decided to return when we assured him that our chameleon would be removed from the premises. He told us that this creature is regarded by some African tribes as a symbol of death and for this reason the people are even afraid to look at it.

In some regions, people erroneously believe that chameleons are very poisonous and if a camel accidentally eats one he will die. The African Bushmen claim, on the other hand, that the chameleon has a miraculous power to bring rain, and pray to it during prolonged drought.

Chameleons have always puzzled naturalists because they have a number of unusual characteristics. They have very peculiar eyes. Located in cone-like turrets, they can move independently, so that one eye can fix on a target in front while the other may be searching for another object below.

Probably the most amazing feature is the chameleon's tongue, which is its weapon and, in some species, is longer than the animal itself. Some chameleon species are up to 60 cm long and their tongues may also reach this length.

The animal has its 'gun' ready to shoot at the first opportunity. When the chameleon spots an insect, it takes aim and, slowly opening its mouth, suddenly shoots out its long tongue at the target — rarely missing. The tongue acts like a released spring and moves with such lightning speed that the human eye only sees a vanishing prey. Only high speed photography can catch a chameleon's weapon in action, which only takes a fraction of a second to strike.

When the prey is hit it is trapped by the tongue's sticky surface and then pulled quickly back into the chameleon's mouth. A large chameleon, such as a Madagascar native, Oustalet's chameleon, (*Chamaeleo oustaleti*) which can be 60 cm long, not only catches large insects but even small animals such as mice or nestling birds.

Chameleons use their tongues as weapons. To become a proficient hunter, the new born baby has to practise for some time. Initially it can only strike accurately from a distance of a few centimetres.

Another characteristic common to all chameleons is their ability to change colour rapidly. There is a common belief that chameleons can change their colour to blend with the objects on which they are sitting. This is only partly true since chameleons can only change their colour within a certain range usually from green through yellow to dark grey. Nevertheless, chameleons have a great talent for disguise as, besides changing colour to blend with branches or leaves, some have developed spiny projections over their bodies which resemble dried leaves or pieces of bark.

Chameleons also change colour with a change of mood. For example, when two males of the African species *Chamaeleo hoehnelli* meet, they do not fight but circle each other threateningly. Soon one of the chameleons' body colour starts to change. From a light green the body colour gradually deepens to become a dark green. In this unusual way one of the contesters shows that he is weaker and acknowledges his defeat without fighting. Strangely, the other male does not change colour at all.

This is rather exceptional, passive behaviour since in most species male chameleons fight fiercely over territory. Chameleons are so aggressive that they may even bite one another during copulation.

The strange ability of chameleons to change colour has become proverbial, and is well known among people in Africa. For this reason in some regions chameleons are attributed with special powers. It is believed that, since a chameleon can change its colour, it can also change human fate and ill health and cure disease. When a Mansa man in northern Ethiopia suffers a severe headache he uses a chameleon as a remedy. The live chamaeleon is placed on his head and it is believed that as soon as the animal changes its colour the pain will immediately disappear.

Mollusc's Harpoon

When a 27-year-old man landed on a small island in Queensland in 1935 the main aim of his trip was to expand his cone-shell collection. He was lucky to find a fine cone specimen and ignored a prickling in his palm, which he felt as he picked up the cone. After some 30 minutes, however, his legs were so weak that he was unable to stand. He finally fell into a deep coma from which he could not be revived.

The deadly poisonous cone was found to be the geography cone (*Conus geographus*). It is very handsome, some 15 cm long and has a unique pattern resembling an ancient map.

Geography cones, as well as other venomous cones, have evolved a sophisticated weapon system which can be compared to the man-made guided missile. Many species of the genus *Conus* manufacture disposable darts which they store in special dart sacs. In bigger species these darts may be some 1 cm long, and they are made of bone and hollow. The cone shoots its darts with the following unique strategy. When it is hungry and prey, such as a small fish or mollusc, is nearby, it fills its dart with venom and shoots it from its long snout, thrusting the dart into the victim's flesh. The poison it injects is so powerful that it paralyses a small victim in just a few seconds. The dart, which is still attached to the base of the snout, is discarded after use and another fresh one takes its place at the tip of the snout so that the cone is ready for the next attack. The venom has a strong paralysing effect on the muscles. When such a dart is used against another snail, it causes it to release the grip on its shell so that the attacking cone can devour the soft parts.

Several cones have such potent venoms that they can kill people. The thrust of the dart is so strong that it can not only penetrate human flesh but can even penetrate clothing. For this reason it is very risky to place an unknown cone in one's pocket. Some people erroneously think that it is only the cone's narrow end which is dangerous to touch. In fact, many venomous cones have very long snouts and the poisonous dart can sting from any part of the shell, including its blunt end. To avoid the horrible stings, heavy leather gloves must be worn, but the best way to avoid the deadly cone's venomous 'bite' is to treat them like poisonous snakes and leave them alone.

The venom of the geography cone can be compared to that of the rattlesnake and a number of people have died as a result of careless handling of these shells. The death rate from the geography cone sting is some 20 per cent of all people stung, which is higher than deaths caused by the cobra or rattlesnake.

The geography cone's sting acts like curare poison by totally inhibiting nerve impulses to muscles and causing paralysis and death in humans. It is, however, an attractive mottled, red-brown and blue-grey shell and is much sought-after by collectors.

There are many poisonous cones and among those deadly to man are the striated coneshell (*Conus striatus*) and textile cone (*Conus textile*). One toxic cone is the famous 'glory of the sea' (*Conus gloriamaris*). This is the rarest and most attractive shell and for over 200 years it was one of the most sought after. A well preserved cone of this species fetched a record price of $2,000 in 1964.

There is an incredible variety of shapes, colours and sizes among some 400 known species of coneshell. Some shells have interesting patterns such as the alphabet cone. This cone, which may be some 10 cm long, is creamy white in colour and is covered with spiral rows of orange or brown dots that strikingly resemble Chinese characters, hence its common name.

Rifle Fish

In 1765 the governor of a hospital in Java, who was an amateur naturalist, noticed the strange habit of a Javanese fish. He saw that this fish knocked down insects resting on branches by shooting at them from the water. He made a detailed report about his discovery and attached a dried specimen of the fish to prove his claim. His account was unfortunately rejected because, by mistake, the governor sent the wrong fish specimen to Europe. He sent a butterfly fish, which the biologists knew lives in the sea where there are no insects to hunt for. Some 130 years later, a Russian scientist studied this fish in an aquarium and confirmed its unusual behaviour.

The fish involved is some 15 cm long and was named the archer fish (*Toxotes* species). The archer fish feeds on water insects, but also seems to enjoy sport and shoots at insects resting on branches overhanging mangrove swamps. The clever fish swims just below the surface and searches for its prey. Once it spots an insect it pokes its mouth up out of the water and takes aim at its target. It then vigorously spits water bullets at the unsuspecting insect, which falls into the water and is caught by the clever artilleryman.

Its shooting accuracy is amazing as it can hit a target from a distance of up to 2 m, which is more than ten times its own length. If the first shot is unsuccessful, the fish corrects its aim and shoots again until the insect falls down. The bombardment is so strong that the jet can travel up to 5 m and if such a water bullet hits a man's face it can cause sharp pain.

The clever fish occasionally chooses a different method. Instead of aiming at the insect it hits the perch and the result is the same. The insect loses its grip and falls into the water where it is snapped up by the waiting archer fish.

The archer fish has the unusual skill of using water as a tool, but gradually has to learn how to aim and shoot properly. Their shots are not accurate at first and they initially only use this technique on a target not more than a few centimetres away. When they grow larger and become stronger and more experienced, their accuracy considerably improves, but they still need a lot of practice before they can shoot with the same precision as their parents.

The archer fish is especially adapted for this remarkable skill. It has a groove in the roof of its mouth and a ridge at the back of its tongue. By raising its tongue, a narrow channel is formed through which the water is squirted. The direction of the 'bullets' is regulated by the shooting fish changing the position of the tip of its tongue. Some archer fishes only fire single shots at a time, while others shoot with the rapidity of a machine gun.

There are some eight species of archer fish which can shoot from the water. They live in both fresh and salt waters from India and South East Asia to northern Australia.

Its unique behaviour has made the archer fish a popular aquarium fish, as it does not stop using its hunting technique in captivity. It is said to occasionally shoot at insects sitting on bulbs above the aquarium, and can give a painful shot in the eye if you stare at it too closely. Some people say that the fish considerably reduces the number of flies and other insects around the aquarium.

Archer fish do not entirely depend on insects captured outside the water as they also feed on small water animals. According to some naturalists, archer fish use their extraordinary weapons because they prefer to eat land insects.

As well as insects and spiders, archer fish may shoot other small creatures spotted on the overhanging branches. In one instance this small fish was reported to have shot down a baby lizard.

Breaker of Hands

People have long been fascinated by the mysterious power of certain fish to produce a peculiar numbing effect when touched. Such fish were called 'narke' in ancient Greece and it is from this word that the present word 'narcotic' is derived. As they could not explain the ability of the fish to cause numbness or shock by mere contact, they believed they had miraculous powers. The species involved was an electric ray found in the Mediterranean.

It was customary to have a numbfish present during childbirth as it was believed to be able to relieve pain. The Romans used the numbfish to cure headaches by placing the live creature on the sufferer's head. Similarly, the eleventh century Arabs exploited the numbfish's mysterious power and used electric catfish to produce small 'shocks' as a kind of electro-therapy to treat rheumatic pains.

Socrates was called a 'numbfish' by his colleagues as he used to numb his listeners 'with the spell of mind-bending oratory' and the fascination with the unique numbfish was so great that it was commonly represented in ancient Greek and Roman mosaics and paintings.

Strange shock-producing fish also occur in other regions. When the first explorers came to South America, they were surprised that local people used horses when crossing even the most shallow rivers. They learned that the people were afraid of the electric eel. This fish (*Electrophorus electricus*), which despite its appearance is not a true eel, is very dangerous to touch. It discharges very strong electric impulses from its body. The American Indians made horses enter the water first because they knew the fish would exhaust its 'batteries' on the horses and they could then safely enter the water.

Rather than using powerful jaws or poison, this remarkable fish has developed an ability to knock down its prey simply by touch. A fisherman once said that while handling such a fish in his boat, he was suddenly 'thrown' by the horrible fish as if he had been knocked down with an axe. Fishermen try to avoid immediate contact with these fish by using rubber gloves and wearing rubber boots.

The larger the fish the stronger is their electric discharge. The electric shock is similar to that which would be obtained if you stuck your moistened finger into an electric outlet, but may be much stronger. A minute, 2.5 cm long, electric eel can make a man's hand tingle when touched, while a bigger fish, say 1 m long, can stun a man. A mature fish is able to discharge electricity of up to 500 volts, and in a few cases a large electric eel discharged a record 650 volts, the greatest ever reported. People have drowned after falling unconscious as a result of the touch of an electric eel.

It is hard to believe that a fish can evolve organs producing electric power similar to a man-made motor-car battery. The electric eel's 'battery' is composed of thousands of electroplates, and an adult eel can have up to 6,000 electroplates, each of which is capable of generating small discharges of about one-tenth of a volt. Usually these electroplates are connected in a series so that their effect is cumulative. The electric discharge is controlled by the fish's brain so that it can use it as a weapon at will.

The electric eel also produces a very weak discharge for navigation and for locating objects. It is a very important ability since an electric eel becomes almost blind as it grows to maturity. The tail, in which electric batteries are located, of a 1.8 m long eel is some seven-eighths of its total length, which means that the main organs such as the alimentary canal, liver, kidneys, heart and gonads are crammed into the remaining eighth of its body.

In Africa, the most well known is the electric catfish (*Malapterurus electricus*), which can attain a weight of some 20 kg. It is capable of discharging up to 450 volts, but interestingly it is only the first discharge which is so powerful and the following discharges are much weaker. This fish, which occurs in West African rivers, is hated by local fishermen. They have a variety of descriptions for this dangerous fish. One which is printable is 'Abubunsamu', used in Ghana, which literally means 'breaker of hands'.

There are many fish which produce electricity, but usually the electric field is weak and does not cause problems to their neighbours. There are some which have recently attracted the attention of biologists. In African rivers there are some peculiar fish called elephant-snout fish (*Campylomormyrus rhynchophorus*), so called because they have strange elongated snouts resembling elephants' trunks. These small fish have 'electric batteries' located under their tails, which produce a weak field. This fish discharges constant impulses and was found to be extremely sensitive to any chemical impurities in water. If such a fish is placed in water containing a minute quantity of impurity, such as phenol, it will detect it immediately as its electrical field changes accordingly. Since it is much quicker to register the fish's reaction than to perform water analysis in the laboratory, it is assumed that these small fish may play an important analytical role in future and help guard the purity of our drinking water as changes in the discharged impulses can be measured by installing special, sensitive apparatus.

18.
Chemical Warfare

Bombardier Beetle

When German scientist Rolander was observing some European beetles in 1750, he was astonished by the unusual behaviour of one of the specimens. He reported: 'When first I took this creature and it straight away gave out a small jet of liquid blue smoke accompanied by a slight report, this unexpected event startled me so greatly that I let the creature fall from my hands to the ground . . . it fired off an explosion, fully twenty times in succession, so that one could only be amazed that so much air should be contained within so small a body.' Small wonder that a beetle with such a remarkable fighting ability has been named the bombardier beetle.

When a bombardier beetle, such as the North American species *Brachinus frumans*, is attacked it uses its incredible chemical weapon to defend itself. A special valve to its storage chamber releases explosive substances into the so-called reaction chamber where they strongly react, forming oxygen. This causes liquid to be forecefully ejected from the anal opening in a fine spray, which is usually sufficient to deter the predator. Interestingly the ejected liquid is as hot as boiling water and is also irritating. The tip of the insect's abdomen is so flexible that the animal can aim with precision to fire at its attacker. The beetle does not fire until it is touched and the sudden ejection of the extremely strong irritant is an excellent weapon as most attackers give up immediately. If necessary, the beetle can fire up to 20 times a minute so its bombarding power is very strong indeed. Although there are several beetle species which can discharge foetid substances from their anuses, none can produce such a vicious liquid jet as the bombardier beetle.

To deter predators, other beetle species discharge poisons. For example, the great water beetle (Dytiscidae) produces a steroid hormone which is released into the water. When a fish swallows the 'poisoned' water it becomes 'drugged' and is soon unable to attack the clever beetle.

One beetle, the ladybird (family Coccinellidae), secretes an amber-coloured liquid when touched which looks like blood. The significance of this 'bleeding' is still unclear, but it is believed that because of its unpleasant smell it has some repellent properties.

Beetles apply an enormous diversity of mechanisms to deter predators, which is not surprising as there are some 350,000 beetle species in the world, from small, 0.25 mm long creatures to giant beetles measuring over 15 cm. While some beetles are very common there are others which are extremely rare. The longicorn beetle (*Titanus giganteus*), which was discovered in 1771, is so rare that only one specimen was found and, despite the efforts of collectors ready to pay enormous amounts of money for one, no other specimen was found for 180 years. Ironically, it was found under a lantern in a small town in north-eastern South America. In fact this species is the largest beetle in the world and it can be up to 20 cm long.

Very large beetles also occur in Africa and are known as goliath beetles (*Goliathus* species). African children occasionally use these beetles as toys. A live beetle is tied to a stick and as it tries to escape it makes a whirring sound like an aeroplane which amuses the children.

Some beetles are famous for their incredible capacity to run over water when escaping from predators. The rove beetle (genus *Stenus*) secretes a surface-active substance from its glands which expands so rapidly that the insect is driven forward with remarkable speed of 75 cm per second over a distance of some 15 m. It has been estimated that this beetle can cover 150 times its own body length in just a second. A motorcar, at a speed of 100 kmh, can only cover seven times its length per second.

Horrible Stinker

Early travellers to America were obviously greatly impressed by skunks, the strange animals which use an unbearable stink to defend themselves. As one writer reported in the fifteenth century: 'It hath a ventositie so strong, and so evil of sent, that it doth penetrate the wood, the stones, and all that it encountereth withall, and it is such that Indians have died with the stench . . . and this smell lasteth fifteene, twentie, or more days, and it is such that if it lighteth neere some towne, it is presently disinhabited'.

Although the above description is a great exaggeration, as nobody has died as the result of the evil smell skunks produce, the fact remains that it is really one of the most unbearable odours nature could invent. The smelly liquid ejected by the skunk can cause temporary blindness if it hits the eyes, causes severe irritation of the nose and mouth and can cause temporary choking.

Although it is exuded by the animal in very small quantities, the liquid has such a strong and lasting smell that it can be likened to the poison gases used in warfare. No wonder the Americans decided to use the skunk's smelly 'gas' in Europe during the First World War. They released some skunk's liquid into the wind blowing in the direction of the enemy lines, and it is said that the Germans immediately detected the smell, concluded that a gas attack was taking place and put on gas masks as quickly as possible. By wearing 'unnecessary', heavy masks their fighting ability was considerably reduced and, by using the 'gas' trick, the Americans achieved quite an advantage.

The skunk's 'gas' is so penetrating that clothing hit by it must be discarded as the horrible stink seems to remain forever.

The stinking liquid has occasionally been used as a medicine and this has sometimes caused problems. When a preacher, who once used this medicine to relieve his asthma, entered his church for a Sunday service, all his parishoners rushed out. The air inside the church was unbearable and nobody could stay in there for even a short time. The preacher's sensitivity to his peculiar medicine was rather low so he was astonished by their sudden response.

One can imagine how strong the smell is when it is 'freshly' ejected by the animal itself. Skunks are real experts at chemical warfare. When it is threatened, it ejects the foul-smelling liquid at the enemy's face. When the liquid hits the predator's face it blinds him for at least two days. The attacker certainly learns a lesson and avoids skunks in future.

The stinking liquid is produced by two glands near the anus and so, to make a good shot, the skunk must present his rear to the enemy. The two glands act as powerful guns which are in the animal's full control. He can fire them both or shoot from one 'gun' only. If he misses his shot for the first time, he can fire again up to six times. But the animal usually tries not to exhaust all its liquid when defending itself, as it then has to wait for a few days before the glands fill again. The skunk can deliver its stinking jet to a distance of about 3 m, but the ejected liquid can be smelt with a good wind blowing, at a distance of more than a kilometre.

The skunk does not immediately use its chemical weapon when attacked. It first gives a distinct warning, and some skunks perform a characteristic handstand before they change their position and start ejecting the irritating liquid. These animals have very distinct, striped or spotted black and white fur, and a large tail which could be considered to be the first of their warnings. It is therefore very easy to recognise a skunk from a distance. They rely on their warning colouration and chemical weapons so much that they do not even avoid cars. Instead of making a quick escape, the skunk shoots his stinking jet at the car and, of course, unfortunately loses its battle.

Surprisingly, the obnoxious liquid of the skunk stink gland can be used to improve certain perfumes. By removing the repulsive components from the liquid, a substance can be obtained which shows a remarkable ability to retain perfume odours for a long time.

Among several species of skunk found in the USA and Canada, the most common is the striped skunk (*Mephitis mephitis*), which bears two white stripes on its back and has a beautiful bushy tail. Its fur is so attractive that, in the past, many thousands of animals were killed every year. The traders initially had problems selling these furs, because people related the furs to the animals' horrible smell. To make the product more attractive the furs were usually sold under the more attractive name, 'Alaskan sable'.

Skunks used to be very common in some regions of the USA. The North American Indians called them 'Sikak', from which the name Chicago is believed to have been derived.

These potentially stinking animals can be tamed and become excellent pets if treated properly. They never use their stinking weapons against good owners and are known to be docile and loving. Usually, however, before the animal is tamed, its scent glands are removed to avoid the possibility of the whole house becoming uninhabitable.

19.
Bizarre Relationship

Candle Thief

Despite its name, the greater honey-guide (*Indicator indicator*) is not interested in the bees' honey but in their wax. It has an incredible desire for wax, but is unable to open the bees' nest to get at it. The clever bird has found a solution to this seemingly helpless situation, however. It asks people for assistance.

When a bird finds a bees' nest in a tree he immediately starts looking for a human in the area. Using its insistent begging call, the bird sends a loud message, as if trying to say that he badly needs human help. Someone usually responds to the plea and the bird then starts to lead him to the nest. It slowly flies in the direction of the nest, stopping frequently and keeping its eye on the man like a real guide, apparently checking the progress of his 'assistant' and incessantly 'begging' him to hurry up.

Once the man reaches the tree containing the bees' nest, the guiding bird falls silent, as if indicating that the trip is over. The man then opens the nest with a machete and, after stupefying the bees with wood-smoke, the happy man gathers the sweet harvest. He usually eats some honey on the spot while the guiding bird patiently waits for his turn. The man leaves some of the combs full of wax for his bird guide. There is an unwritten law in Africa that some of the comb must be left for the honey-guide. It is believed that if one refuses to do so and fools the bird, it will take revenge and next time it will not guide someone to a nest, but to a place inhabited by a ferocious forest beast which will kill him.

Interestingly, this curious partnership between a bird and a man, which must have been practised since time immemorial, remained unknown to the scientific world until recent times. It seemed so improbable that biologists, who learned about it from early travellers, regarded it as a typical, amusing traveller's story. It remained a mystery until 1955 when Herbert Friedman, an American ornithologist who studied the behaviour of honey-guides for a long time, reported his results.

He described many occasions in which he was guided by the birds and confirmed that the honey-guide relishes wax, but admitted that it also eats the bee larvae.

If a man does not cooperate with this clever bird, it will search for an entirely different partner. This time he looks for a forest animal which is only fond of honey and does not like bees' wax at all. This new partner is the honey badger (*Mellivora capensis*), whose skin is so tough it is impervious to bee stings.

The honey badger also quickly recognises the characteristic call of the honey-guide. He allows his partner to guide him to the nest and their cooperation is almost identical to that between the human and the bird. On reaching the nest, the honey badger opens the nest with its strong claws and eats the honey, leaving the bird plenty of broken combs beneath the tree. Africans are aware of this relationship and whenever someone hears the bird's characteristic call for assistance, he quickly tries to forestall a honey badger who may be around.

Honey-guides have a fantastic sense of smell and are able to detect the presence of wax-containing materials from a long way off. The smoke from burning wax is immediately recognised by the bird. When we were staying in Kenya we were told a story of a priest who was unhappy because his wax candles often disappeared from the altar. He suspected his parishioners, but later discovered that it was a bird who was stealing the candles.

'Imprisoned' Wife

Among the most spectacular birds found in tropical Africa and South East Asia are hornbills, so called because of their fantastic and enormous bills, often adorned with large horny casques. This remarkable bird attracted the attention of the first European travellers who took back many tales about it. It was believed that the male had a strange habit of imprisoning his female partner in a nest-hole and that, while feeding his imprisoned wife, he had no time to feed himself so that he eventually died.

There is some truth in these stories since almost all hornbill species feed their partners while they incubate the eggs sitting inside the nest-hole, walled-in as if they were prisoners. But the female's imprisonment is self-imposed and even the choice of nest must be accepted by both partners. The breeding habits of the silvery-cheeked hornbill (*Bycanistes brevis*), which lives in the forests of East Africa are the best known. These birds make a lifelong partnership, and, after mating, both partners search for a suitable hole in a tree to construct a nest for their offspring. When the female accepts the location, she enters the hole with the aim of plastering up the entrance to close herself inside the nest. It is team work, however, and the male brings mud as building material which he regurgitates to the female through the entrance. She keeps plastering the entrance with the mud pellets so that after a few days only a narrow slit is left. The self-imprisoned female lays her eggs in the nest and starts incubating them, while her partner supplies her with food. The female accepts the food through the slit and also removes faeces and other refuse from the nest to keep it tidy.

The male partner is an ideal husband. He seems to work non-stop gathering food, mainly fruit, and delivering it to his imprisoned wife. It has been estimated that a male can deliver some 24,000 fruits to his wife and then to his family during the $3^1/_2$ month period of imprisonment. When the youngsters hatch after some 45 days they remain with the mother for about eight weeks and then the mother breaks the 'prison' wall to set them free. The small babies are ready to fly the moment they are freed from the nest.

Some hornbill mothers escape first, while the youngsters are left alone to mature. In this case, the babies are now fed by both partners. Once the mother has damaged the nest wall by escaping, there is a danger of predators attacking the defenceless young. But once the female escapes the youngsters immediately start repair work. From their faeces and saliva they make their own building material and are able to close the entrance almost completely. When their flight feathers have grown, the babies break the 'prison' wall and join their parents waiting outside.

Ground hornbills (such as the *Bucorvus leadbeateri*) is an exception to the rule. In these species the female is not 'imprisoned' at all. The birds live in a colony and incubate their eggs in a 'normal' nest which they build together. Strangely, their colony, which consists of some eight individuals, is not governed by a male but by one of the females. She sits on the eggs laid by all the female birds, while the other members supply her with food.

There is a strange belief related to ground hornbills in some parts of Africa. The corpse of the dead bird emits an offensive odour and they believe that if such a bird is thrown into an empty river bed, the river will fill with water again. The river, they say, will become sick, and by trying to wash the stinking bird away, it will invite the rain to fill it.

Among hornbills, perhaps the most spectacular is the rhinoceros hornbill (*Buceros rhinoceros*). It is, in fact, the largest hornbill (1.5 m long) and the enormous casque on its bill resembles a rhinoceros horn hence its name. This fantastic bird even used to be worshipped in Malaysia, and it is the emblem of Sarawak state. It occurs in Borneo and Sarawak and the people are proud of this unique bird, calling their country the Land of Hornbills. The bird is a great attraction for both biologists and ordinary tourists.

One of the largest hornbills is the helmeted hornbill (*Rhinoplax vigil*) which can be up to 1.3 m long and occurs in Malaysia and Indonesia. Unlike other hornbills, this giant bird has a solid casque which is so heavy that, together with the skull, it makes some 10 per cent of the bird's total weight. These bills became a source of ivory for making ornaments and thousands of these beautiful birds have been killed to supply the merchants. In the past, the ivory attracted Chinese merchants who used it to make belt buckles, which were popular among Chinese high officials, especially during the Ming Dynasty, from the fourteenth to the seventeenth century. The hornbill's ivory, which was called 'ho-ting' used to be more valuable than elephant ivory.

Boxing Crab

The most bizarre of the many species of crab is the fiddler crab (genus *Uca*), although it is only the male which is so unusual in appearance. Its right claw is so incredibly large that it can be almost as large as the whole crab. It is as if he has borrowed a claw from another giant crab.

Although one would imagine such an enormous claw would be very awkward for its owner, it in fact plays an essential role in his life. It is its main organ of display for attracting a member of the opposite sex.

The giant claw is usually brightly coloured and is used in combination with leg movements to produce a sort of semaphore code. Each crab species has its own particular and individual pattern of giant claw movement. In one species the claw is moved like a violinist's arm, hence the common name of the whole genus, fiddler crab. In other species the enlarged claw waves up and down as if doing 'press-ups' while in some the claw moves outwards and upwards in three distinct jerks. There are some 70 species of fiddler crab which mainly occur in mangrove swamps, and each of them has its own specific display. Since many similar-looking crab species often inhabit the same area, living together in huge crowds, without specific display it would not be easy for a male to attract a female of his own species.

Male fiddler crabs are known for their remarkable sense of beauty. Sometimes, the arrival of an especially attractive female can cause considerable confusion among waiting males. At a particularly 'sexy' female's appearance males who have already chosen their mate abandon their first choice and turn to the more attractive new 'lady'.

The crabs also direct their signals at other males to mark their territory. If a signal is disregarded a fierce fight may break out. It is a short fight in which each crab pats the other with its giant claw. They interlock their weapons until the claws seem to be at breaking point, but after a short time the weaker crab accepts his defeat and retreats to his burrow.

While the majority of crab species have strong shells which protect them from predators, some species, such as the so-called hermit crab have no shells. These crabs have found a clever remedy to their problem. They simply use the shells of gastropods as their own shelters. They may spend much time searching for a suitable empty shell and may be seen patiently waiting for a particular gastropod to be eaten and so vacate its shell. Even then, the hermit crab 'thinks twice' before using it. It examines its size and shape carefully and then finally accepts it as his new 'apartment'.

Certain hermit crabs, (*Eupagurus prideauxi*), have also 'discovered' another method of defence against predators. They invite another animal to defend them by establishing a peculiar association with sea anemones which are known for their deadly stinging tentacles. A crab finds an anemone which he attaches to the back of his shell. In this way the crab has constant protection against many of his predators while the anemone has a free ride on his back to the areas where food is more abundant. So the relationship is beneficial to both partners.

When a growing crab's shell becomes too tight and he has to change it, the anemone must be detached. Once the hermit crab has chosen its new home, the same anemone is immediately attached to the new shell and continues to serve its partner as before.

Certain crab species, such as the attractive tropical crab *Lybia tesselata* use anemones in an entirely different way which is also very strange. In order to frighten its enemies, the crab holds an anemone in each of its claws and uses them like a boxer, hence this crab's popular name, boxing crab. In this case, however, it is not the strength of the blow inflicted but the anemone's stinging tentacles which matter. By thrusting it's 'gloves' at them the boxer crab deters predators which usually retreat. The crab also uses its anemones for hunting. The crab kills small fish with the anemone poisons by punching them with its gloves. The fish now contains the anemone poison, but the crab seems to be immune to it and is unharmed.

The boxing crab may modify his technique and hold both anemones in one claw so that the free claw can be used for capturing small creatures. Its claws have special projections so that the anemones can be firmly held together. When holding two anemones together in one claw, the 'boxing glove' may be as big as the crab itself.

Physician of the Sea

While most small fish panic and quickly escape when they spot a large and dangerous fish, there are some which seek their company. Surprisingly, when a big, dangerous fish meets one of these small fish it does not attack it but instead is evidently 'pleased' by the meeting. The secret behind this rather unusual behaviour is that the large fish need a service which can only be provided by small fishes.

These small, 'brave' fish are known as cleaner fish and they are able to remove parasites found on large fish. Their teeth protrude like pairs of forceps from protractile mouths, which enable the cleaner fish to gently and quickly pluck the parasites from the fish's skin with a tweezer-like action. They really take care of the fish as, besides removing skin-irritating parasites, they also remove irritating debris and clean wounds. Their suffering patients are obviously rather grateful for such service and never attack them.

The best-known cleaner fish is the common cleaner wrasse (*Labroides dimidiatus*), a swift and agile fish with distinctive blue and black stripes. They usually work in small groups of two or three and have a permanent 'cleaning station', usually at the depth of about 3 m. These cleaning fish seem to understand the role of advertising in their profession as they perform a kind of graceful dance in their station to make themselves better identifiable to potential clients.

Large fish are said to form a queue to wait their turn for the cleaning service. According to Limbaugh, who observed some cleaner fish behaviour, up to 300 large fish were serviced by one cleaner fish in just six hours.

The cleaning is both efficient and fast. It is a rather extraordinary sight to see a huge fish with its jaws gaping like a patient in a dentist's clinic and a small fish not being eaten but serving its needs. Small cleaner fish not only enter and spend some time inside the fish's large mouth, but also enter its gills to remove parasites there.

Even if a large fish has its mouth almost closed while being cleaned, it leaves sufficient space for its 'guest' to leave safely after its job is over. It has been estimated that most large fish require the service of cleaner fish at least once every two weeks.

When a fish of the *Disceus* species meets a cleaner fish it gives it a special signal to attract its attention in order to be cleaned immediately. The large fish changes its skin colour almost black to show the cleaner fish the whitish patches infected by the irritating fungi.

Interestingly, various fish species meet at the cleaning stations, but although they would certainly attack each other outside the 'station', here they behave like gentlemen and do not attempt to hurt one another.

Cleaner fish seem to be so obsessed with cleanliness that they occasionally try to 'clean' other unwilling creatures met in the water. One diver said that he was once frightened by a small fish which started plucking the hairs on his leg. The cleaner fish must have been searching for skin parasites.

Surprisingly, certain fish, which are not cleaners, must have observed the relationship between cleaner fish and deadly sharks, since they have exploited this behaviour to their own advantage. An aggressive fish, *Aspidontus taeniatus*, has developed the blue stripes characteristic of some species of cleaner fish. In this disguise and by imitating the cleaner fish's dance it can even approach a large and deadly shark. But instead of picking off parasites, the wicked fish bites off a large piece of flesh and darts away to the complete surprise of the wounded shark.

Some cleaner fish, for example the already mentioned common cleaner wrasse have other interesting habits. They live in large groups of up to sixteen females, which are kept in line by one male. This male is king and defends his harem from predators. The females are curiously competitive and one of them is always dominant. If the king of the colony dies, this dominant female replaces him. Amazingly, within some two hours of the disappearance of the male ruler, this female miraculously transforms into a male. This is not just a change of duties, but a transformation in a full biological sense and she changes her sex and becomes a male. The new king will, of course, show the same aggressive attitude as his predecessor towards the female members of the harem. When the newly created male dies, the next dominant female is always available, as if simply waiting for her chance. In such an unusual way, the colony is always under the rule of a male king — very strange arrangement indeed.

20.
Insidious Creatures

Blood Suckers

In the heart of Sri Lanka used to be the wonderful Sinhalese kingdom of Kandy. Although in the early sixteenth century the Portuguese controlled all the coastal areas of the island, they could not conquer this wealthy kingdom. Their attempts to penetrate the jungle were unsuccessful and, the story says, their troops were not defeated by humans but by a powerful army of blood-thirsty leeches.

Although most leeches live in water, one species (*Haemadipsa zeylanica*) found in the tropical forests of Sri Lanka live on the leaves of the forest shrubs and trees. The so-called terrestrial leeches occur in many regions, especially in South East Asia. As they feed on human or other animal blood, they have to wait for occasions to suck blood. They have a powerful sense of smell and quickly detect an approaching animal or human and, congregating in huge numbers, attach themselves to a passing host. They go for the uncovered parts of the body, usually attacking the ears, neck and even eyelids and then start sucking blood. They secrete a substance with their saliva which stops blood from clotting and 'suck' until their guts are full. It may take 20 minutes before the leech gorges itself and drops off and its body may swell to five times its original size.

It is not surprising that millions of these leeches, which could literally cover the soldiers' bodies, stopped the Portuguese penetrating the jungle. Those who lost their way in the jungle, the story says, were later found hopelessly lying on the forest floor unable to defend themselves from these blood-thirsty creatures. Many people have died from being attacked by leeches in the forest. It is believed that in Sri Lanka in the sixteenth century only a few, specially dressed, Portuguese Franciscans were able to travel through these dangerous leech-infested forests.

It is hard to believe but many people in Europe used to voluntarily allow leeches to be attached to their bodies to suck blood. The leeches used are common in European rivers and ponds and they became known as the medicinal leech (*Hirudo medicinalis*). Their use has an interesting history in Europe. It was initiated by a French physician, Francois Broussais, who claimed that any disease from whooping cough to madness, could be cured by using leeches. He explained that the sucking creatures would draw out with the blood all the 'evil vapours' responsible for the disease. His power of persuasion was so strong that doctors started to use leeches in their daily practice. It was quite normal, in those days, for a doctor to carry a jar full of 'miraculous' leeches with him when visiting his patients.

Some 1,600,000 litres of blood was sucked from the French people by medicinal leeches every year. To supply the medical profession with leeches enormous numbers had to be collected from local waters. It is said that during the leech craze in France, 45 million were gathered in 1893. When the leeches in France became almost extinct many millions were brought from Hungary and Russia. As a result, they soon became extinct in many European countries. Special farms were then established where leeches were grown for medicinal purposes.

Although there is no evidence that blood letting, or 'leeching' as it used to be called, is a better method of treatment than others such as using medicines, there are still people who believe that 'leeching' is most effective in lowering high blood pressure. Although leeches are now considered an 'obsolete' medicine, strangely, in recent years, there seems to be a revived interest in some of their medicinal values. When a plastic surgeon performs a skin graft, the tissue often becomes heavily congested with blood. It has been found that applying leeches can effectively reduce the swelling.

The study of how leeches keep blood flowing led to the discovery of a powerful anticoagulant called hirudin. This substance, which can be isolated from leeches, has become an important medicine used for the treatment of certain heart conditions.

Leeches are in fact very interesting animals. They are very easy to keep as they can live a remarkably long time without food. One chance to suck blood is sufficient for them to live on for six months. They have a remarkable storage capacity and their digestion takes place gradually and there are unconfirmed reports that some leeches can survive after a single 'big' meal for up to 18 months. Biologists explain that once a leech finishes the supply of blood, it slowly consumes parts of its own stomach and it can survive on this unusual food for a few additional months.

The medicinal leech, which is a few centimetres long, sucks up to 15 ml of blood, which it stores in special pockets in the gut. Here the blood is dehydrated and thickened and is not attacked by bacteria. So the leech has developed a technique to survive a long time on a single meal, which is an excellent adaptation for an animal which can only eat occasionally.

Some leeches can attain a great length. A real giant among the many leech species is the giant South American leech *Haementaria ghilianii*, which can be up to 50 cm long. Scientists became interested in this species because it produces comparatively large eggs, which are useful for research purposes and particularly for the study of the development of the nervous system in lower animals.

Not all leech species are blood suckers as some feed on small animals such as worms, insect larvae or snails and some are scavengers. Some leeches (*Haemadipsa* species) in tropical Asia, which live on land, are dangerous to animals in another way. They can enter animals' respiratory tracts and make breathing difficult.

Blood Thief

Probably no other animal has provided the subject for so many horror stories as the tiny bats called vampires. Frightening stories were told of a giant vampire bat who drained blood from helpless seamen at night. The best known vampire is the ferocious character, Dracula, who changes himself into a human vampire to drink the blood of his human victims. In the past, blood-thirsty vampires were thought to be the corpses of the dead who rose from their graves at night to drain blood from innocent, sleeping people.

It used to be believed that before the vampire bat begins to suck blood he lulls the victim by spraying its nose with a soporific scent. This is not true, but the bat's strategy is still quite interesting. Once it finds a sleeping victim such as a horse or a cow or a man, it lands carefully nearby and then stealthily creeps in search of a bare part of the body such as an ear or a toe. When attacking a sleeping person, the bat usually chooses uncovered toes to drain blood. The vampire's teeth act like small razors and do not inflict a wound which would alert the victim. Instead, the bat chisels away a small slice of skin in such a gentle way that the sleeping victim is not wakened. The seeping blood is then lapped up. In the bat's saliva an anticoagulant is introduced which prevents the blood clotting. This secures a free flow for a long time until the hungry bat is full. He is usually so gorged with his easy meal that he is unable to fly away for some time. The clever blood thief is known to return to the same victim night after night to suck its blood. It has been found that vampire bats specially locate their colonies near herds of animals to have a steady supply of blood. It is common for a vampire to pick out the same animal from a large herd.

Although a single vampire does not 'drink' much blood, a colony of a hundred common vampires can drain a total amount of blood equal to that of 14,600 chickens or 25 cows. It has been estimated that the common vampire (*Desmodus rotundus*) takes only some 30 ml of blood at a time, but with thousands of bats involved, thousands of litres of blood may be drawn from domestic animals in just one year. One vampire bat in captivity consumed 23 litres of blood in one year.

The amount of blood lost by a human sucked by a single vampire is small and does not inflict much harm. But if a small child is chosen by a vampire bat, the free blood flow caused can lead to death. As the bat introduces a powerful anti-coagulant, the bleeding may continue for several hours after the bat's visit, and this may have severe consequences as the loss of blood may be substantial.

The vampire's 'bite', although very delicate, is sufficient to transmit rabies. In the West Indies such vampire-bat-transmitted rabies led to considerable losses of domestic animals.

It seems strange that the vampire bat's night visit does not wake up its victim. The bat's whole blood stealing operation is so wonderfully stealthy that even if a man was expecting such a visit, he would not be able to 'feel' the presence of the unwanted visitor. Fortunately, there are only three species of vampire and they are in Central and South America. The remaining bat species are harmless because they destroy a large number of harmful insects.

21.
Merciless Killers

Ferocious Ants

It is hard to believe that in some tropical regions people leave their homes and run in panic when they learn that they are about to be invaded by army ants. Their behaviour is reasonable, however, since the huge colonies of ants march like organised troops, destroying all living creatures on their way.

No matter how big the creature is, it will be devoured by the ants if it cannot escape in time. It is said that ants devoured a horse which could not be released in time, leaving only its skeleton. Small animals may be stripped to the bone in less than a few hours. Like a black avalanche, the ants move across the village, entering every corner of the house, climbing the ceilings and entering the drawers and cupboards. Even with windows and doors firmly closed, the ants find ways to get inside buildings.

In tropical Africa, the army ants, often called driver ants (*Dorylus* species) march in columns, usually flanked by soldiers with huge jaws who keep the rest of the army, which is composed of blind worker ants, in order. Such an army may be composed of up to 20 million individuals. The first indication of an approaching column of army ants may be a flock of birds swooping over and trying to eat them. The native people immediately set all their domestic animals free and move out of the path of the marching ants until they pass through. The only advantage of such an invasion is that the hungry ants free homes of rats, cockroaches and flies and thus eradicate pests for some time.

Horrific encounters with army ants have been experienced, for instance by the famous explorer of Africa Dr David Livingstone. An army ant 'force' attacked him while he was asleep in his tent and hundreds of ants literally covered his body. He managed to get out of the tent but his suffering continued for a long time. His men removed the ants from his body one by one and even after they had finished their laborious work, it took several days before the explorer recovered from the horrid effects of the attack.

Even more frightening are the army ants in South America,

particularly Brazil. These are the ants of *Eciton* species, which are commonly known as legionary ants. According to one report, a legionary ants army advanced on a small town in Brazil in an enormous column, covering an area 1.5 km long and 800 m wide. This invasion caused real panic among the inhabitants who asked the local police to take action. During the confrontation with the hordes of marching ants one policeman lost his life. Despite this sacrifice, they were unable to stop the terrible ants.

Various methods, even fire, are tried to stop the invasion of such 'armies', but in vain. The swarms of ants continue over the burnt bodies as if crossing a bridge. Even if the marching hordes meet a river they do not change direction. They immediately form a bridge with their own bodies over which the remaining troops pass. These ants can also survive floods. They form a compact ball of bodies, which float on the water. Although the ants on the bottom of the ball drown, the great majority survive the flood.

In some parts of South America the painful bites of the ferocious legionary ants have been utilised as a cruel form of punishment. The victim is placed in a cage which is then placed in the path of a marching colony. One can imagine the horrific torture experienced by a victim who cannot escape the ferocious ants.

The army ants' bite can be exploited for the benefit of people. In some regions, these ants have been used for 'surgery'. The edges of severe wounds are brought together and the ants are placed one after another to bite the wound edges together. The grip of their powerful jaws makes a perfect suture. Once the ants have performed their job, their heads are cut off so that only their jaws remain in place until the wound has healed. It is said the technique is very efficient and the wounds quickly heal.

The ferocious army ants are also sometimes useful in another way. When the separatist war broke out in Nigeria (Biafra) a great famine devastated the local population. An incredible number of people saved their lives with the 'help' of army ants. As a last resort, they gathered these ants to feed themselves and especially their starving children. This unusual food was a real blessing for many people and they have even been called 'sausage flies' to emphasise their value as a food. Recent studies have shown that they contain high amounts of protein and are highly nutritious.

Army ants have no permanent nest or home. Instead, they form a living nest. Hooked one to another, millions of them form a huge ball. The living nest, made of their own bodies, even has chambers and in the centre there is a room for a queen and her brood. Some army ant species construct their living 'nests' on tree branches, which makes them hard to detect among the green leaves of the tree canopy, so the colony is well camouflaged and protected against predators.

Butcher Bird

The shrike is a small, peaceful-looking bird and there is nothing striking about its pleasant carolling song. But this bird is one of the bloodiest killers, and his only instrument of massacre is his very sharp, hooked beak. This bird is an expert in aerial manoeuvrability and can hover like a falcon, selecting its prey from the air, even attacking a bird bigger than itself. After catching its prey, it impales its victims on thorns or barbed wire and thus seems to imitate a butcher in his shop, hence its common name, the butcher bird. He often only eats one of his victims and leaves plenty of impaled animals for later. Because of this habit of impaling its victims, the bird gained a bad reputation for being exceptionally cruel, but this is not true, because the bird kills his prey with his strong beak before impaling it.

Some ornithologists accuse the bird of killing excessively and say that it kills just for fun. In fact, the bird fills its larder to feed its young or when food may be scarce. It may seem that the bird kills more than he and his family can consume. One butcher bird, the Mexican loggerhead shrike (*Lanius lodovicianus*), makes a large collection of small lizards impaled on thorns and then seems to forget about it. But when lizards become scarce in the region because of a severe drought, the clever bird returns to his larder and eats the previously killed lizards.

Shrikes have amazingly well developed sight and their vision can be compared to that of a man using binoculars, so it can easily detect a 'suitable' prey from great heights. Because of their remarkable skill and agility, these birds were used for hawking instead of falcons. In fact, King Louis XIII kept shrikes in cages ready for hawking at his palace.

An Australian bird (genus *Cracticus*), also known as the butcherbird, has the nasty habit of impaling its prey live. Strangely, these cruel birds are among the best singers. A pair of these birds is known to perform a kind of duet in which each bird sings alternately. The male starts by singing one phrase while the female immediately replies, using exactly the same tune; it is difficult to tell whether it is one or both birds performing.

While butcherbirds are harmless to humans, although notorious killers of small animals, there are some birds in Australia which are much more aggressive and strike at people even when unprovoked. These birds are called butcherbirds by some people, but are, in fact, Australian magpies which are unrelated to the common European magpies. Although Australian magpies (*Gymnorhina tibicen*) generally leave people alone, their attitude changes during the nesting period when they may attack anybody approaching an area where they are nesting. The bird unexpectedly swoops from above and often from behind and may hit the person's head with its strong beak. A magpie once struck a girl in the eye seriously damaging her vision.

In some areas where magpies are common, children are not allowed to leave their classrooms unless they wear protective hats and magpie nesting areas are considered real 'danger zones'. During magpie 'terrorism', which lasts some three months of the year, people are advised to wear sunglasses to protect their eyes outdoors.

It has been noticed that in some areas magpies seem to hate school boys in particular as they ferociously attack them, while often completely ignoring the girls. Some biologists try to explain this behaviour by saying that the birds are taking their revenge. Since some school boys often throw stones for fun and disturb the nesting magpies, the birds, which have good memories, easily recognise their enemies and attack them long after the naughty boys have forgotten the incident.

Living Scissors

The most vicious and voracious fish of the Amazon River basin are certain species of piranha such as the red piranha (*Serrasalmus natteri*). These relatively small fish have razor sharp teeth and may bite anything that moves in the water. No wonder that the first European travellers to the Amazon River basin brought back incredible stories about piranha attacks. According to one popular story, a man on horseback was devoured by dreadful piranhas. He was riding through a river in the Amazon region when he fell into the water and was attacked by piranhas with such great ferocity that, shortly after, only his skeleton and that of his horse were found in the river. The man's clothes were said to have remained intact.

Many people wading the rivers in northern South America have been attacked and severely injured by shoals of piranhas. For instance, during his travels to Guyana, Sir David Attenborough, a famous British naturalist, met Tiny McTurk who said: 'I remember once getting into a canoe with fifteen Indians. We got in one at a time, and in doing so we had to put one foot in the water. None except me had boots on . . . It turned out that thirteen out of fifteen had small pieces of flesh bitten clean out of their feet.' In fact, in some parts of Guyana, nearly everybody living near the rivers has lost his or her fingers or toes when bathing or crossing a river.

According to one report, a tragic accident occurred in the piranha-infested Urubu river in the Amazon region of Brazil. A bus full of passengers tumbled into the river, but when the rescue team arrived several hours later, most of the 38 passengers were said to have been reduced to skeletons by hungry piranhas. In 1981 Brazilian newspapers reported that more than 300 people were devoured by piranhas when a cargo-boat sank at the Brazilian port of Obidos.

Piranhas' jaws are well equipped for their job. Their mouths are equipped with exceedingly sharp, triangular, flat teeth with which the fish can chop off a finger like a piece of carrot. Fortunately, their mouths are not big and can usually only cut out a small chunk of flesh. Sharp piranha teeth are used by some Indian tribes as razor blades, and they call normal scissors 'piranhas', to stress their sharp edges. Piranha teeth are also used to make their curare-poisoned arrows more effective. By attaching them to their arrows the wounds inflicted are bigger and the penetration of the deadly poison is more rapid. In some regions the Indian warriors adorn their masks with piranha jaws to make them more frightening.

Piranhas are cunning hunters. They keep very close to their potential victims for a long time, pretending a complete lack of interest, and then, with lightning speed, they attack the unsuspecting prey. Piranhas usually do not work alone, but in large shoals which may be composed of many hundreds of fish. As the prey starts to bleed, piranhas appear as if from nowhere and devour the prey with unusual ferocity. Even a large animal can be stripped of all its flesh in just a few minutes. There is a confirmed report that a large, wounded alligator was completely stripped of its flesh in just five minutes by a piranha shoal. The frenzy of a piranha feast is so unusual that when particularly hungry they may even devour their own kind.

These fish are even dangerous after being caught when handled by fishermen. They might look completely dead but can still be alive and have been known to bite off a finger or a toe of an unwary fisherman.

The ferocious nature of piranhas has been exploited in some areas of the Orinoco region for an unusual purpose. As these regions are often flooded for several months of the year, it is impossible to bury the dead in graves and so the piranhas 'help' to solve the problem. The dead body is simply thrown into the water and hungry piranhas devour all the flesh, leaving only a clean skeleton. The skeleton is then taken out of the water, dried, painted and decorated and then decently buried.

In some rivers of Brazil, piranhas are so abundant that fording a stream with herds of cows is a real problem. Sometimes an animal is deliberately dumped in the river to attract piranhas to this particular spot. All the piranhas from the area (sometimes thousands of fish) rush to share a big meal while a huge number of cattle can cross the river unmolested, just a few metres away.

Although piranhas are normally carnivores, they sometimes change their feeding habits and become vegetarians. This happens when a large area of forest becomes flooded and piranhas find themselves in water-covered areas where animal flesh is not available. The fish then start to eat fruit which fall from trees and these become their normal food, a change of feeding habit which surprised researchers. Some piranhas, however, never feed on animal flesh and are born vegetarians.

Pirates of the Sea

The magnificent frigatebird (*Fregata magnificens*) has an enormous wing span of over 2 m and has been likened to a frigate, a fast sailing ship, hence its common name. It is also known for piratically attacking and robbing other seabirds of their food, hence its other common name, the man-of-war bird. It seems to be constantly at war with other birds, although it does not cause them much harm. Soaring high above the sea or hanging almost motionless in the air, the 'pirate' looks for birds carrying fish and, with its remarkable agility and speed, it swoops down and attacks its victims. The clever pirate can easily distinguish a seabird full of fish from one flying 'empty'. Having spotted his victim he overtakes it, forcing the frightened bird to drop down or disgorge its catch of fresh fish, which the flying pirate catches and eats. It is a remarkable aerobatic display, since the frigatebird can snatch the falling fish with remarkable accuracy and never misses.

There are five species of frigatebird, which live on the coasts of tropical seas and they all have the same piratical habits. They have specialised in robbing such birds as boobies and pelicans, especially during the breeding season. They have no problem catching any seabird because they are the fastest flying seabirds. The record held by one individual bird was 154 kmh in still air. Amazingly, the ground speed of this bird is even faster. Ground speeds of over 320 kmh have been reported.

These birds do not rely solely on robbing other birds of food as they also catch fish, even dead ones floating on the surface. They usually attack fish near the water surface and flying fish in particular, because they cannot swim and their feathers are not waterproof. The short period of a few seconds when the flying fish is in the air is sufficient for the acrobatic bird to snatch it with ease. These birds are so adept that they can snatch small twigs from calm water without even making a ripple.

The birds' extraordinary aerobatic abilities are not hereditary. The young are taught this skill from their parents and undergo a long training during which they learn to swoop after falling leaves or twigs. Being able to catch objects in the air is such an essential skill that youngsters who do not become experts in this technique may not survive when their parents stop feeding them.

Despite their skills frigatebirds are still dishonest creatures. When adult birds find young seabirds of other species alone, they swoop and snatch them up for food. The pirate birds are known to locate their nests near colonies of other nesting birds so that they can rob them when opportunities arise.

Frigatebirds usually nest in large colonies in trees on small islands in tropical and subtropical seas. These colonies can occasionally become unforgettable spectacles. During the breeding season all males of the colony suddenly change their appearance. Each has a throat sac which is normally orange in colour but to attract females he inflates his pouch like a balloon which is several times bigger than the bird's head. The balloon turns scarlet. During his display the male frigate looks quite extraordinary. To advertise his presence and to attract a female he rattles his beak and spreads his wings. Once a female is lured, copulation takes place on the ground and the pair then starts to collect twigs for the nest. The male's flying capacity during the breeding period is only slightly reduced even though it has to 'carry' its enormous balloon, but the sight of such a bird is really unforgettable.

This bird, although somewhat infamous for its piratical habits, can become a good friend to humans. The inhabitants of the Polynesian islands have always admired their flying ability and even their habits and on Easter Island it was even worshipped. In some regions they became tame and were regularly fed with fish by the local inhabitants. The tamed birds became so friendly that, even when set free, they would later return from their ocean trips to the same owners.

These birds have even become postmen when normal postal services were unavailable. Using tamed frigatebirds, messages could easily be exchanged among the inhabitants of remote islands. The technique was simple. A letter or a small gift was inserted into a reed cylinder and attached to the bird's feet, and the bird would then fly with it to a distant island. The messenger bird landed on specially constructed perches and the 'letter' would be delivered. This method of letter-sending was very efficient and early missionaries used the 'air-mail' service to keep in contact with other missionaries living in the remote Polynesian islands.

Aborigines living on some Australian islands call frigatebirds 'rain brothers'. There is a belief that whenever these birds come ashore and stay, heavy rain will soon follow.

22.
Perilous Stingers

Lady of the Beautiful House

With its ferocious appearance and powerful sting at the end of its segmented tail poised for immediate attack, the scorpion (order Arachnida) has always had a bad reputation.

In ancient Egypt, the goddess Selket was depicted as a lady with a scorpion on her head. Called the 'Lady of the Beautiful House', this goddess was responsible for the protection of another god who was in charge of people's embalmed entrails. In the New Testament the scorpion was regarded as the symbol of evil people who did not accept Christ's teachings. In medieval Europe the scorpion became a symbol of Judas and the sign of betrayal and dishonesty. Pliny claimed that the scorpion's sting is always fatal to virgins and usually deadly to married women and widows, but not to men. According to an ancient myth of the Mexican Toltecs, their fertility god was changed into a scorpion when he could not resist the temptation to have an affair with the goddess of love. As late as the seventeenth century it was believed that a man born under the sign of Scorpio would become a murderer.

Despite their very bad reputation, however, in some regions scorpions have been used as a cure. It was believed that when a dead scorpion was applied to a fresh wound it would heal it rapidly. Because of a general fear of their often deadly sting, scorpions have also been used in another way. Among members of the Dervish sect in Tunisia, 'conquering' a live scorpion became a strange ceremony. A man who wanted to demonstrate his bravery had to swallow a live scorpion with its venomous tail. Successful demonstration of such bravery gave the man holy status in the community.

In some regions of Somalia people believe that having the carved image of a scorpion protects them from being stung when asleep. Many people die from scorpion stings around the world. In Mexico alone, some 2,000 people die each year as a result of scorpion stings. Scorpions are very insidious as they enter tents or huts at night and sting people while they are asleep. A tragic case was reported from Pakistan. Nine men were sleeping in a primitive country guesthouse and they were all stung by scorpions during the night. Only one man survived.

During the Second World War, soldiers fighting in some desert regions of North Africa were less afraid of their human enemies than of scorpions who attacked them in their tents at night. Many soldiers died in this way, without ever fighting a battle.

Among the most deadly is the fat-tailed scorpion *(Androctonus australis)* which lives in deserts of North Africa. Its poison is so powerful that it can be compared to the venom of deadly snakes such as the cobra. This scorpion has a fat tail and is known to attack people without provocation. Its sting is lethal to a healthy person if the antivenom is not available in time. These scorpions are especially dangerous to barefooted children.

There are over 800 species of scorpion and the virulence of their venom varies a great deal from one species to another. Only about 50 species are a serious threat to humans and scorpions living in Europe and Australia only inflict stings which are painful, but rarely fatal, even to small babies. To attack, the scorpion curves its tail over its head and drives the sting firmly into the victim's flesh. The swollen bulb contains enough venom to inflict several stings one after another when the scorpion is stepped upon or molested.

Scorpions are solitary animals and hate the presence of other scorpions, except during the brief mating season. If two scorpions are put in a cage together they will fight to the death, and the winner devours its foe. When scorpions fight one another they do not use their stings, only their claws. The fighting animals seem to realise that their venom would not kill their rival. It takes 200 times more scorpion venom to kill a single scorpion than is necessary to kill one guinea pig. There is a common belief that when a scorpion is surrounded by a ring of fire from which it cannot escape, it will commit suicide by stinging itself to death. This, however, is not true.

Some scorpion species have interesting courtship rituals. The male grabs the female of his choice with his pedipalps and walks her backwards, forwards or even sometimes sideways and their tails intertwine. The 'dance' can take several hours, but, at a certain stage, the male deposits a packet of sperm on the ground and pulls his partner backwards until her sexual opening takes up the sperm.

Scorpions do not lay eggs but give birth to living young as the eggs develop inside the mother's body after fertilisation. The scorpion babies, which may be up to 30 in number, are exact copies of their parents. The very small newborn babies are all pure white and as soon as they are born they immediately climb onto their mother's back. A mother covered with white babies looks very strange indeed. The babies remain there for some time, protected by the mother's powerful stinging tail, despite the fact that they themselves can inflict an effective sting as soon as they are born. When the youngsters moult they change colour and the mother simply drops them off her back and they start independent lives.

Strangely, if their mother is very hungry, she may even eat some of her babies during the early stage of their development.

Contrary to a common belief, scorpions do not live only in desert regions, and can even be found near the Himalayan snow-line. Scorpions are extremely resistant to heat or cold and can live frozen in a block of ice for some time. They can live without food for almost a year and without water for three months. Some scorpion species are only fully active when the temperature reaches 45°C, which is a very unusual adaptation.

Scorpions are among the most ancient animals since they are known to have existed 400 million years ago. No wonder that they have a fantastic ability to adapt to the harshest conditions and can survive when most other animals would perish.

Sea Wasp

The world's most dangerous marine stinger, the box jellyfish (*Chironex fleckeri*), which occurs in shallow sea waters of Northern Australia, was unknown to scientists for an incredibly long time. Despite the fact that this creature was responsible for more deaths than sharks or crocodiles in Australia, for a long time biologists were sceptical about its existence.

Although some local doctors were aware of a deadly kind of jellyfish, they hesitated to mention it when describing the cause of a death. They preferred to use other reasons for death, such as an 'allergic shock' due to a certain marine creature of unknown origin.

The authorities did not show much concern about a marine stinger which could not be described more precisely, but with the outbreak of the Second World War the situation changed. Due to the threat of a Japanese invasion, military bases were established on the Northern Territory shores, where thousands of Australian soldiers and sailors were trained. It was a shock to the community when, long before any invasion, soldiers were eliminated from training by an 'unknown marine stinger'. Not only were several soldiers killed by the deadly stinger when swimming in the sea, but many had to be hospitalised. Efforts to catch the strange killer were unsuccessful and were soon abandoned.

Only when a 5-year-old boy died on the beach in northern Queensland in 1955, was a live specimen of the stinger discovered. The jellyfish specimen was sent to Dr Ronald Southcott, an Adelaide zoologist, who was the first to examine this unknown marine animal and who gave it its scientific name.

This box jellyfish is composed of a nearly square 'bell' the size of a man's head with long tentacles on each of its four corners. They have up to 16 tentacles on each corner, each some 3 m long.

The bell of the box jellyfish is very pale blue and so transparent that it is almost invisible, especially if the water is disturbed. The ribbon-like tentacles are covered with minute stinging cells, called nematocysts, containing a deadly venom. Some tentacles exude adhesive substances, which enable them to stick to live objects in the water. These nematocysts are extremely tiny and are so densely packed that up to 1,000 venomous spines can inflict a sting from an area no larger than a pinhead. It has been estimated that one box jellyfish can have a thousand million stinging nematocysts and so even if the tentacles touch only a small area of skin, severe poisoning occurs. If the total length of attached tentacles is more than 6 m, death may occur in just one minute. It is regarded as the most venomous marine creature in the world.

A person suffers excruciating pain when stung, comparable to the effect of a hot iron on the skin. Although the victim may lose consciousness, he may still be revived if the resuscitation technique is applied immediately. The inflicted part of the body is covered in linear wheals which develop into red blisters and the scars may remain after recovery.

For some reason this creature's sting was compared to a wasp's sting and it was called a sea wasp. This common name for the box jellyfish has caused confusion, especially among visitors. When they hear about the dangerous wasp in Australian waters, they nervously search the sky and ignore the marine creatures. There was a story of some people who panicked when they were bitten by a 'normal' fly while bathing in the sea off northern Australia.

This deadly box jellyfish occurs in the waters of northern Australia for only about 3 months of the year and the public is warned by special posters. Parents are encouraged to teach their children not to touch even dead sea wasps they may find on the beach, as they can still inflict a serious sting.

Bluebottle

Sailors are sometimes startled when they see hundreds of tiny 'ships' moving across the tropical seas. This unforgettable sight is caused by the Portuguese man o' war (*Physalia physalis*). Although its common name indicates a single animal, it is in fact a colony of modified polyps and medusae.

The small, bluish 'ship' on the sea surface is one of the modified medusae, while the rest of the colony is totally submerged. This medusa is an unsinkable float, inflated with a gas produced by an internal gland, which is a mixture containing up to 13 per cent carbon monoxide, a highly toxic gas to humans and higher organisms. This float provides the buoyancy for the whole colony which travels with the sea currents.

The submerged polyps are specialised to perform various functions for the benefit of the whole colony. While some are responsible for reproduction, others perform digestion and the most elaborate are those which catch food. These are the so-called 'fishing tentacles', which can reach a length of up to 10 m.

These food-catching tentacles are covered with millions of stinging cells called nematocysts. Each nematocyst contains a tightly coiled thread-like tube which discharges its venom with remarkable speed like a syringe when touched. As the victim, usually a fish, struggles, more and more stinging cells are activated and the numerous tiny injections of poison paralyse the fish. When this is accomplished the tentacles move the prey closer to the mouth where it is eaten, digested and absorbed by other specialised polyps.

The Portuguese man o' war is called the bluebottle in Australia and feeds mainly on small fish, which become entangled in the deadly tentacles. It is rather unusual, therefore, that some fish are on 'friendly' terms with the bluebottle. There are at least two species of small fish (*Numeus gronovii* and *Schedophilus maculatus*) which are known to live among the tentacles and are fully protected from other predators in this unusual shelter. It is still a mystery how these fish can do this and it used to be assumed that they were immune to the bluebottle's venom, but recent studies suggest otherwise. It is now believed that they have acquired acrobatic skills which enable them to avoid direct contact with the stinging nematocysts. It is still unclear why the Portuguese man o' war does not catch and devour such an easy 'meal'. According to some biologists, the host benefits from their presence. They think that the distinctively coloured and attractive fish simply lure other, bigger fish into the trap.

Bluebottles are hazardous to humans, although lethal poisoning is extremely rare. Nevertheless, if a swimmer or diver touches the stinging tentacles, the sting is usually severe. A person being stung usually thrashes about in the water and cries for help as the sensation is like a cigarette burn. The pain is usually severe and radiates in all directions. It usually lasts 1–2 hours, depending on the number of tentacles touched and red welts like lashes appear on the skin. The shock from an unexpected Portuguese man o' war's sting can be so intense that it causes fainting and it can lead to heart failure and even death in people with weak hearts.

It should be noted that children are the most vulnerable to the sting of bluebottles. Since up to a hundred of these attractive, bluish balloons can be washed onto the beach, they may attract children's attention who pop the inflated balloons to enjoy the delightful explosion. But if a balloon is stepped upon barefoot, the tentacles beneath the 'bottle' can still be dangerous. Even a bluebottle which has been on the beach for days and appears to be dead may still retain its stinging ability, so it is better to leave it alone.

Although the tentacles of the bluebottle frighten most small animals, there are some which are not afraid of them and even devour them. Such brave animals are small molluscs called nudibrachs (*Glaucus* species) which are known to feed on the 'deadly' tentacles, despite the nematocysts. What is even more surprising is that these molluscs are able to 'extract' the stinging cells from the tentacles and somehow transfer them into their own filaments. In this unusual way the mollusc itself acquires a stinging capacity, but it is unclear whether it just uses them for defence or also to attack prey.

Death from Curiosity

When two friends went spear fishing one day in 1954 in the sea near Darwin in the Northern Territory, they did not expect a tragic end to their trip. While they were fishing, one of them noticed a strange bluish creature and picked it up out of curiosity. He then tossed it onto his friend's arm where it stayed for a while before dropping back into the water. When they came ashore the man on whose arm the animal had crawled for a while, felt a dryness in his mouth which was followed by difficulty in swallowing. His condition deteriorated so fast that the other man decided to take him to the nearest hospital. On the way, however, the victim's breathing ceased completely. His body became blue and despite medical efforts to revive him the man died. It was established that he had been stung by a small octopus called the blue-ringed octopus (*Hapalochlaena lunulata*). This marine creature, which is not more than some 20 cm across, is among the most deadly Australian sea animals. It does not attack humans and is harmless if left alone, but if taken out of the water it gently makes a small incision in the skin that can go unnoticed by the victim. Through a small parrot-like beak the animal injects its powerful venom which is so strong and acts so fast that there is usually not enough time to help the victim. The poison of this octopus quickly blocks the commands of the human brain which results in paralysis of breathing and death. One 'bite' may be sufficient to kill an adult.

The blue-ringed octopus uses its powerful weapon for hunting. When it is hungry it bites into an animal's back, usually a crustacean, paralysing it so that it is unable to defend itself and devours it. The animal does not like using its poison sting unnecessarily. If it is not really hungry and just wants a bit more food, the clever octopus applies another technique. It glides over the victim's body and instead of biting the animal it squirts some of its toxic substances into the water close to the prey. It waits until it is paralysed by the venom and then feeds on it.

Unlike other octopus species, which use an ink screen as a defence against predators, this octopus has no such ability. It does not, in fact, need this kind of defence trick as its powerful poison, one of the most deadly poisons ever provided by nature, is sufficient.

This dangerous marine animal has a peculiar breeding habit. After mating, which may take several hours, the female lays up to 200 eggs, which are only some 7 mm accross, and carries them fastened to her tentacles. The mother's devotion to her eggs is extraordinary. For the whole incubation period, which may take two months, she constantly washes the eggs by squirting water from her siphon. She is so busy washing and aerating her eggs that she has no time to feed herself. When the eggs hatch nobody is there to greet the youngsters as their mother starves to death.

There are two species of highly dangerous octopus. While the *Hapalochlaena lunulata* is confined to waters off the Northern Territory and Western Australia, another closely related species (*H. maculosa*) is common in subtropical waters from southern Australia to southern Queensland. The latter species is smaller and is no more than some 12 cm in length. Both animals are deadly to humans. They both have the characteristic faint blue circles on their surfaces which darken when the animal is disturbed and become a very distinct blue. The best protection against these deadly creatures is to avoid contact with them when spotted.

It should be noted that blue-ringed octopus temporarily occur in some lakes adjacent to the sea as is the case in the West Lakes in Adelaide, South Australia. Dangerous blue-ringed octopuses also occur in the seas off Papua New Guinea, New Zealand, Fiji, Indonesia and Japan.

Crown of Thorns

The inhabitants of the Ryukyu Islands of Japan have suffered many invasions throughout history. But, in recent times, the most threatening invasion was not human, but that of a mysterious sea creature. Local fishermen complained that they were unable to harvest seaweed because they said that whenever a fisherman entered the water on the coral reef, he was stung by a sea-snake, which resulted in agonising pain. Scientists soon found that the source of the sting was not a sea snake but a starfish. Their enemy was in fact the crown of thorns starfish (*Acanthaster planci*) which occurred there in large numbers.

The crown of thorns looks quite deadly. From its central disc it has twelve to nineteen 20 cm long arms which are all covered with long sharp spines, which may be up to 6 cm long. These spines inject poison, when trodden on or touched, causing acute pain. It can cause fainting, nausea and vomiting and even temporary limb paralysis. No wonder that divers and swimmers hate this prickly sea creature and call it the devil starfish.

The crown of thorns can be up to 60 cm in diameter and occur mainly on coral reefs, because they eat calcareous coral polyps, which seem the most unsuitable food imaginable. The way the crown of thorns eats the hard polyps is most peculiar. It pushes its stomach through its mouth and attaches it to the polyp surface, gripping like a vacuum cleaner. It exudes digestive enzymes which liquify the hard polyp tissues. The animal has such a strong appetite that in one night it can eat coral reef equal to half its own body size. After many hours only the coral's bleached skeleton is left and the starfish moves on to look for another victim. An adult crown of thorns can consume up to 12 square metres of coral a year, and millions of such animals infesting the coral reef can destroy the coral formation over an extensive area.

In recent years, the crown of thorns starfish has started to endanger the Australian Great Barrier Reef because some 14 per cent of the Reef is heavily infested. About 20 square kilometres within the Arlington Reef was completely destroyed by the crown of thorns in just three months.

It is a very difficult enemy to control, because of its enormous reproductive capacity. A mature female can spawn 20 million eggs and its hatchlings float on the surface and easily infest new areas. Even if large numbers of eggs or hatchlings are eaten by predators, their population can increase dramatically.

With their spiny bodies, the crown of thorns should have no enemies at all when mature, but there are some marine animals which attack and eat them. Among its few enemies, the most important is the large mollusc called the giant triton shell (*Charonia tritonis*). The giant triton uses its powerful foot to flatten the spines of the animal and then sucks out its soft tissues with its proboscis. Some biologists believe that giant tritons could be a means of controlling the crown of thorns population and that they should be protected. These huge tritons are in fact prized by shell collectors, many of whom profit by producing various ornaments, such as lampshades, from the shells.

Another enemy of the crown of thorns, which feeds on its flesh, is a shrimp (*Hymenocera picta*). To make their attack successful, these shrimps work in pairs. The two shrimps first press on the animal's tube feet so that it loses its grip and then turn it onto its back to expose its flesh to be cut off and devoured.

Despite their enemies, millions of these dangerous crown of thorns still infest the coral reefs of the Pacific region. Biological methods of control used on an experimental scale have not been very encouraging.

It has now been assumed that only more radical measures such as complete elimination of the starfish from given areas can bring good results. The quickest and most efficient method would be to employ an 'army' of divers to inject a poisonous solution of formaldehyde into one of the starfish's arms which kills the whole animal. Unfortunately, this method would be too costly and would be impossible to apply on large areas of coral reef.

The Japanese have chosen a simpler technique and have achieved very promising results. They have collected the starfish one by one from the sea bottom, taken them to the surface and destroyed them on land. In some of the Ryukyu Islands local fishermen collected over 6 million crown of thorns. It took months of very hard work performed by thousands of people, but due to their efforts the whole coastline in the region was totally freed of the spiny devil.

Marine biologists in Australia are aware of the danger of this animal to the Great Barrier Reef, and some scientists believe that once the coral reef is destroyed, it will never regain its former beauty and a great, natural heritage will be lost for ever. Conservationists argue that if financial support from the government to control the crown of thorns population is not sufficient to save the Great Barrier Reef, it would be Australia's greatest act of negligence in history.

Warty Ghoul

One of the most extraordinary fish has to be the stonefish. It lives on the sea-bottom and looks like an irregular piece of stone rather than a fish, hence its name. Its whole body is covered with warts and its skin is brownish-black in colour so that its disguise is almost perfect. As it lies motionless at the sea bottom amid the debris of a coral reef, it is almost impossible to detect. When a barefoot swimmer steps on a stonefish, he receives a nasty sting, yet is usually unaware of its source. The pain is so excruciating that a victim may become delirious and lose control and may even attack those who are trying to help him. One man rolled on the ground and then tried to cut off his infected leg. The shock caused by a stonefish sting may cause death from heart failure. In lesser cases the agonising pain persists and, if medical help is not provided, the limb may become gangrenous. Morphine is usually used to relieve the pain, since stonefish are among the most venomous fish in the world. The horrible stinger is also called the scorpion fish or 'warty ghoul'.

The fish does not deliberately attack humans, however, and simply responds quickly when disturbed, using its well developed defence system. Along its back the stonefish has 13 dorsal spines, each surrounded by a pair of glands full of venom. These spines normally lie flat along its body but, when the fish is disturbed, they become erect. When pressure is applied the poison shoots from the glands through the spines like hypodermic syringes. It has the largest venomous glands of all fish.

Stonefish discharge their venom with an almost explosive power and the 'syringe' can even penetrate the sole of ordinary sandals. In the shallow, coral reef waters, people become easy victims of the horrible stinging fish. A woman was once seriously stung even though she was wearing shoes — the sting 'found' a small hole in her shoe. The most vulnerable are those who think they will be able to see the dangerous creatures, but do not realise how excellent the stonefish's camouflage is. Some authors say that walking barefoot in the water where stonefish can be found is like walking through a minefield, and this is not a great exaggeration.

Sometimes people step on stonefish and do not suffer the usual horrible effects, but escape almost unharmed with only slight skin injuries. The reason for this 'miracle' is that the stonefish has already exhausted its poison supply by stinging other creatures. It needs about two weeks to recharge its poison glands with new poison.

Interestingly, even a dead stonefish is not safe to touch. Contact with its spines may still cause a so called 'passive sting'.

The stonefish is well known to Australian Aborigines who hate it. As they usually walk barefoot when catching fish on the coral reef, they often become their victims and call them sea devils. Young Aborigines learn to avoid this horrible fish and in parts of North Queensland it is involved in their traditional rituals. During their initiation ceremonies, they imitate the horrific effects of contact with this fish, but use a wax model of one rather than a live fish.

When 'really' stung, Aborigines use various, often surprising, remedies to dull the terrible pain. In Arnhem Land they use the juice exuded from the bodies of cockroaches of the *Cosmozosteria* species which have been heated over a fire. Stonefish venom loses its toxicity when exposed to dilute acids, and therefore fishermen in Queensland first used hot vinegar to treat stings. Fortunately, stonefish antivenom is now available in Australia.

The stonefish does not sting to attack, only to defend itself. Nonetheless it is an excellent hunter. Lying well camouflaged on the sea bottom, it carefully watches its surroundings. When a fish or prawn comes near, the stonefish rapidly snaps it up and although it appears motionless, the stonefish opens and shuts its mouth with lightning speed.

There are three species of stonefish, namely *Synancea horrida, S. trachynis* and *S. verrucosa* and they occur in the Indo-Pacific region, most often off the coasts of Australia and South Africa. They usually grow to about 30 cm in length and can live for many hours out of water.

23.
Living Fossils

Ghost out of the Past

One of the most extraordinary reptiles is the tuatara (*Sphenodon punctatus*), which resembles a lizard but is not one at all. It has the skull of a crocodile, the ribs of a bird and some features of its brain and its mode of locomotion are similar to those of an amphibian. Despite its strange features, for the first 36 years after its 'discovery' it was regarded as an ordinary lizard. It was first scientifically named by Dr Edward Grey in 1831 and it was only in 1867, when another specimen reached the British Museum, that a zoologist named Dr Albert Gunter realised that .the creature was a living fossil. We now know that the tuatara is the only living remnant of an entire reptilian order, the Rhynchocephalia, which appeared 220 million years ago, well before the appearance of the first dinosaurs.

Tuataras live on some 20 small islands off the coasts of New Zealand in extremely cold, wet and windy conditions where any modern reptile would not be able to survive. The tuatara has the lowest active temperature of all reptiles, between 6° and 13°C, whereas in all other reptiles activity begins at a minimum of 14°C.

The male of this peculiar animal, which is up to some 60 cm long, has a soft, spiny crest on its back and tail resembling a comb from which comes the Maori name 'tuatara', meaning 'peaks on the back'. Tuataras played an important role in Maori rituals and beliefs. The Maori goddess of fire, associated with tuatara, was believed to have slain her own grandson, which resulted in misery for the whole world.

The tuatara is described in various Maori legends. According to one, a young man named Ngararanui, abducted a beautiful girl from his tribe and went to Rotorua where they got married. The woman disliked Rotorua and insisted that they return home. She tried to obtain forgiveness for her husband's misdeed, and praised him by claiming that he was so brave that there was nothing in the world he would be afraid to do. In response to her boasting the community elders decided to teach them a lesson and gave him a live tuatara which they ordered him to eat. The man did as he was told and ate the abhorrent reptile. This act of bravery made such an impression, the story says, that all his sins were forgiven and he became a national hero. Since then, he has always been represented on carvings as a man with a tuatara emerging from his mouth. The tuatara thereby became a symbol of bravery in Maori tradition and has also become a symbol of fertility. It is represented in various Maori sculptures and ships are decorated with an emblem depicting the tuatara.

The tuatara is an extremely slow animal, and spends nearly all its time either asleep or in a state of half sleep, falling asleep even when eating. It moves very slowly and its general metabolism is extraordinary — when resting, the strange animal may not breathe for up to an hour. The tuatara seems to be indifferent to pain and shows no distress even with broken ribs and limbs.

Another peculiar feature of this animal is that the male has no copulatory organ. In order to pass semen, the male and the female have to press their urogenital openings firmly together. But this is not the end of the surprise. The tuatara has a strange, vestigial, 'third' eye which can be seen on the top of its head when it is very young, but in older animals it becomes covered over, and invisible. Although this tiny eye, which is about 5 mm in diameter, has a lens, retina and nerves, it cannot focus or form a picture and so cannot function as an eye. Its presence remains a mystery.

Probably because this animal has such a low metabolic rate, it has a higher life expectancy than any lizard or snake. In fact, tuataras can live up to 150 years. It reaches sexual maturity at twenty, but does not stop growing for the first 50 years of its life.

The female deposits her eggs nearly a year after mating and incubation takes about 15 months which is the longest incubation period among reptiles. Young tuataras, which are some 10 cm long, take many hours to break their hard egg shells.

Tuataras often live in burrows made by petrels or other birds. They feed on the insects which themselves feed on the birds' guano. It is said that the reptiles sometimes eat the friendly birds' eggs and may even devour their chicks. The tuatara, the living fossil of New Zealand, is now under strict protection, but, despite this, its numbers have dramatically diminished in recent years.

Discovery of the Century

On 22 December, 1938, Miss Latimer, the curator of a museum in South Africa, arrived to inspect some fish brought in by one of the trawlers. She did not expect to make a discovery of the century. She was really astonished when she saw a strange, bluish fish about 1.5 m long, with peculiar fringed fins on fleshy stalks. Miss Latimer decided to write to

Professor J. L. B.Smith, an eminent fish specialist, and ask him to come and examine the most extraordinary fish she had ever seen. When Professor Smith saw the fish he exclaimed: 'I would hardly have been more surprised if I had met a dinosaur on the street'. Scientists had thought that this fish had been extinct for about 70 million years. He named it *Latimeria chalumnae* in honour of Miss Latimer and to commemorate the Chalumna river near the mouth of which this incredible fish was caught.

This fish had only been known from the study of fossil materials and was believed to be the oldest fish. Specialists believe that vertebrates evolved their legs from its fins. In fact, this fish which is called the coelacanth, is the most ancient living animal with a backbone.

Unfortunately, when Professor Smith arrived, the fish's body was so decomposed its internal organs had to be removed and discarded. In order to make more detailed study of this remarkable living fossil it was of paramount importance to find another specimen. To help the search, special pamphlets, including a picture of the famous fish, were distributed along the coasts of southern and eastern Africa. A reward of £100 sterling was offered to anyone who could catch such a fish and report the find. The poster read: 'Look carefully at this fish. It may bring you good fortune. Note the peculiar double tail, and the fins. The only one ever saved for science was 5 ft (150 cm) long . . . If you have the good fortune to catch or find one DO NOT CUT OR CLEAN IT IN ANY WAY but get it whole at once to a cold storage or to some responsible official who can care for it, and ask him to notify Professor J. L. B. Smith of Rhodes University, Grahamstown, Union of S.A., immediately by telegraph.'

Despite the enormous publicity, nobody could collect the generous reward for almost 14 years. It seemed that no other fish of this species had survived until in 1952, a fish which fitted the description was caught by a modest fisherman, some 1,500 km from the first discovery. It was caught by Ahmed Hussain near the Comoro Islands between Madagascar and Tanzania. The observant fisherman notified Professor Smith, who arrived by plane to inspect the catch. He was able to confirm that it was a second living specimen of coelacanth. It was then revealed that this fish, so important to scientists, was in fact well known to some of the Comoro Island fishermen, but they were very rarely caught because they live at a depth of at least 150 m. The islanders did not actually like the famous fish's taste, and only used its very rough skin as a substitute for sandpaper.

After the second catch, several other specimens of coelacanth were also caught, including one pregnant fish. This last catch confirmed that coelacanths give birth to live youngsters, which had already been extrapolated from studies of fossil specimens.

Studies of the internal organs of this unique fish have revealed many other peculiarities. They are arranged in a very unusual order. Its kidneys, for example, do not lie 'normally' under the backbone, but on the floor of the abdomen and instead of being a pair of separate organs, they are joined together. The coelacanth's backbone is made of hollow rods of cartilage, hence its common name — coelacanth, meaning 'hollow spine'. Its heart was also found to be very simple in structure, much simpler than the heart of 'normal' fish and was exactly as postulated by palaeontology studying fossil remains.

The coelacanth holds another world record. Its eggs are the size of tennis balls and are therefore the largest of all fish eggs. A 1.6 m female was found to contain 20 such eggs.

24.
Bizarre Birds

Flower Kissers

Some hummingbirds are so tiny that the first Europeans to arrive in North America thought that they were not birds but large insects. The stories they brought back to Europe were ridiculed as the fantasies of travellers. Even when the skins of these minute birds were shown to the people at home, they were regarded as native Indian forgeries and were said to have been produced to make a profit. Hummingbirds certainly are exceedingly small when compared to other birds. The smallest is the bee hummingbird (*Mellisuga helenae*), of Cuba, which is only about 6 cm long and weighs 2 grams.

No other bird in the world, not even the lovely bird of paradise, surpasses the hummingbird in beauty. Many hummingbird species have iridescent plumage of almost every colour combination and often look like shining rubies. Some naturalists call these birds the 'gems' of the avian world.

The ancient Mexicans were interested in these birds' fantastic feathers and used them to adorn their clothes. The emperor and his court decorated their mantles with hummingbird feathers. The most popular were the feathers of the *Calothorax lucifer*, a hummingbird whose beautiful iridescent colour and striking appearance earned it its name, the lucifer hummingbird.

Even in recent times, the inhabitants of some regions of South America have worn tiny hummingbirds around their necks to attract partners and their feathers are still used to decorate clothes. In some regions, dried hummingbirds are worn as earrings. Haitians believe that these beautiful birds can bring happiness and luck and in some regions love philtres are made from the dried birds in a belief that they enhance male sexual performance.

They also used to be in demand in Europe for making artificial flowers and even dusters and thousands of birds were slaughtered for this purpose and shipped from America to Europe. The tiny birds would have been completely exterminated but fortunately, just in time, their export was finally prohibited.

It is rather surprising, but the tiny hummingbird is one of the best flyers in the avian world. Only hummingbirds can hover in one spot for several minutes, literally suspended in the air. The smaller birds beat their wings up to 80 times per second, the fastest rate of all birds. Only the helicopter can be compared with this unusual bird's ability, which is also unique in that it can not only fly up and down but also backwards with remarkable ease.

The birds' hovering capacity is very important because they feed on flower nectar. Since many flowers are very fragile and cannot support even such a tiny bird, the only way they can feed is to remain suspended in the air. By hovering in front of the flower the bird sips the sweet nectar using its long beak and highly specialised, long tongue. No wonder Brazilian Indians call these remarkable birds 'flower kissers'. The shape and size of the beak usually suits the type of flower the particular species of hummingbird visits. The sword-billed hummingbird (*Ensifera ensifera*) has a thin beak which is as long as its head, body and tail combined, and perfectly suited for reaching nectar from certain long, tubular flowers.

Hummingbirds feed almost without a break for some 14 hours a day and, in its frantic search for nectar, it may visit up to 2,000 flowers, extracting their contents which, apart from nectar may include small insects.

These birds can also fly very fast and reach up to 65 kmh in a straight flight.

One particular hummingbird likes to cover long distances. The ruby-throated hummingbird (*Archilochus colubris*) makes a 9,000 km trip from Panama to Canada and back, and flies non-stop some 960 km of that journey, crossing the Gulf of Mexico. To make the trip the birds use fat stored in their tiny bodies, which can be some 50 per cent of their total body weight. No other hummingbird has such a fat storing capacity, or makes such a long journey

It is not easy to catch a hummingbird. Augusto Ritchi, a Brazilian biologist, invented a peculiar method. Despite their tiny size these birds are very brave and are known to be aggressive even towards hawks ten times their size and even attack their own mates. He exploits their aggressive nature and uses a pigmy fowl as a decoy, the enemy of hummingbirds in the region. By placing a decoy on a tree and imitating the characteristic call of the pigmy fowl, he attracts the hummingbirds he is looking for. He patiently waits under the tree and when the hummingbirds arrive he traps them, using a long aluminium rod, smeared with a sticky oil. The tiny birds are trapped like flies on fly-paper. This method does not harm the tiny hummingbirds, who can be carefully detached and used for his studies.

The world's smallest birds have always attracted ornithologists and as a result almost 320 species of hummingbird have been identified. Some of these species are so rare that they were only known to exist because some skins had been exported from South America to Europe. For example, the Loddiger's racket tail hummingbird was only known to science by a single skin brought to Europe in 1840. It took 40 years before the bird was finally discovered in a small valley in the Andes.

Bird of a Thousand Voices

The extraordinary Australian lyrebirds remained unknown to the scientific world until 1797. Early Europeans learned of their existence from an infamous ex-convict who caused constant trouble living among the Aborigines. One day he was arrested by the police who found him clothed in a kangaroo skin. By telling them about an unusually beautiful bird which they had never seen, the man was pardoned. As a result, the authorities organised a special expedition west of Sydney where they found the famous bird which was later called the superb lyrebird *(Menura novaehollandiae)*.

This bird occurs in southern Victoria and south-eastern Queensland and is famous for its remarkable tail of unusual size and beauty, which the male bird displays during courtship. A. H. Chisholm vividly describes its transformation 'Slowly, majestically, as though responding to the pressure of a secret spring, the great tail began to rise and spread. The two large outer feathers extended to each side, displaying a beautiful silver-white background . . . two long central feathers, grey, slim, gracefully curved, rose and were held at an angle above the back. A wealth of fern-like plumes . . . formed a silvery fan for a moment, and then descended, like an elfin parasol, over the back and the head. What an astonishing transformation was this! A plain brown bird, through the simple act of raising and spreading the tail so as to display the undersurface, had become beautiful beyond words'.

The shape of this amazing tail, which can measure up to 60 cm, strongly resembles a Greek lyre, hence the bird's common name.

Standing in his display position on top of his mound, the male bird sings his melodious love song, which can be heard a kilometre away. This bird is an incredible vocalist, able to mimic other bird songs and also to repeat any sound or tune he hears. Amusing stories are told about this strange habit of the superb lyrebird. According to one, a surveyor in Victoria was annoyed one day when he saw through his binoculars that his team had started to work without him giving them the usual signal. He reprimanded them for their insubordination, but the workers claimed that they had clearly heard his whistle. It was later found that the order to begin work was given by a lyrebird.

In another case, the extraordinary vocalist disrupted the system of signals used by timber workers in a Victorian forest. Because of the great distances between teams, special signals were used for various events, particularly emergencies. Three whistle blasts meant an accident, while six whistle blasts signified death of a mate as a result of an accident. One day the workers heard six whistle blasts, so they ran, tormented by the apparently tragic event. To their surprise there was no fatality or injury. Apparently, a clever lyrebird had heard three whistles before, but for some reason had added another three to its repertoire.

This bird is even known to imitate such sounds as a car horn, and some farmers in Victoria claim that they imitate the loudness of a quarry at full blast'. It can mimic calls of birds which sing duets, such as whipbirds, or even the difficult 'laugh' of kookaburras. The male superb lyrebird is also known to mimic the barking of dogs.

Another Australian species, the Albert lyrebird (*Menura alberti*) is also an excellent mimic. The male's tail is similar to that of the superb's but without the lyre-shaped outer tail feathers. This bird is said to mimic the car horns, train whistles and even the sound of a horse pulling a heavy wagon. Both lyrebirds can imitate the sound of a violin or piano and their repertoire is so wide that they are known as the birds of a thousand voices.

Unfortunately, the early European settlers were more impressed with the lyrebird's extravagant tails. Year after year thousands of them were slaughtered for their tails, which were sold overseas for ornamental purposes. Although some efforts were made to reduce the exploitation of these birds, as late as 1911 it was said more than 800 lyrebird tails had been exported to Europe from a single NSW dealer that year. Fortunately, lyrebirds are now under strict protection, but despite this their number has diminished in recent years, because of the destruction of their forest habitat. It should be said that fortunately several sanctuaries have been established in Australia for the famous birds.

God of the Air

When the Spanish conquistadors arrived in the Aztec empire they were impressed by the clothes of the Montezuma nobles which were richly decorated with the emerald-green feathers of an unknown bird. They soon learned that these beautiful feathers came from a magnificent bird called Quetzal, meaning 'god of the air' and that the Aztec god, Quetzacoatl, was a serpent with feathers of the famous bird.

The quetzal (*Pharomachrus mocino*) is one of the most beautiful birds in the world. About the size of a dove, it has a spectacular, long tail of metallic-green plumes, the central pair of which extend some 60 cm. Its head is ornamented with a distinct tuft of green feathers.

These beautiful feathers were only worn by explorers, nobles and high priests. Ordinary men were strictly prohibited from using them and if anyone was found collecting them or killing the sacred bird he was often punished by death. Thus the famous bird was not hunted by local inhabitants without a special permit. Those who were allowed to catch these birds did so without causing them harm, only removing the precious plumes from their tails. Only the male birds have beautiful tails and the females were not hunted. In order not to harm the birds the Indians shot from a distance, using a special, gentle technique. Instead of using poisoned arrows they used soft clay pellets to knock the birds over. When Hernan Cortés visited Montezuma's court, he was presented with a special gift. It was a gold blowpipe, used for 'hunting' quetzals, from the emperor's rich collection.

Unfortunately, the Spanish conquistadors and those who came later did not use the innocent blowpipes as it was much easier to kill the birds for their precious tails. However, scientists who now collect these birds for their studies usually use the same gentle method of shooting at them with soft clay bullets.

The quetzal ranges from southern Mexico through the forests of Central America, but is becoming increasingly rare. The bird is the national emblem of Guatemala and the second largest city in Guatemala is called Quetzaltenango in its honour. The highest honour of this country is the Order of the Quetzal and the main monetary unit used in Guatemala is named after the bird.

The male bird helps his partner to incubate their eggs in the nest they make together in a hole in a tree. There was much controversy as to how the male could sit on the eggs without damaging his fantastic long tail. Native Guatemalans claimed that the quetzal nest had two entrances so that he could settle down on the eggs and leave his tail dangling outside the nest, eventually leaving the nest by the other 'door'. This, however, is not the case. The male enters the nest and then turns around to face the opening, folding his long tail over his back so that only the tip hangs outside, so protecting his precious tail.

Some people criticise nature for creating such an uncomfortably long tail, but in fact it improves the bird's stability in the air. The male quetzal's aerobatic skill is so good that it can pick fruit from a tree from the air.

Stink Bird

The baby hoatzin (*Opisthocomus hoazin*) is one of the most peculiar birds in the world. A newborn hoatzin is featherless and can be likened more to a prehistoric reptile than to a bird. What is unique about this young bird is that it has two claws on each of its wings, unlike any other living bird. Only the most primitive bird, the Archeopteryx, which is known from fossil materials, has powerful claws on its wings. No wonder that biologists initially regarded the hoatzin to be a link between modern and prehistoric birds.

In fact these claws help the youngsters escape from predators. When in danger they can crawl along the tree branches on their four 'feet'.

Their nests are usually on branches of trees over water, and the youngsters are even known to escape into the water. They are excellent divers and swimmers and can quickly cover a good distance, fooling the predator. When the danger is over, the baby bird clambers back up the tree using its claws like a lizard, and returns to its nest.

It is rather unusual for young birds to be able to swim before they have learned to fly. When they lose their claws after a few weeks they also lose their ability to swim. Although the adult bird has no traces of its childhood claws, it seems to remember having had them. They continue to use their wings as if they want to grip onto branches with their former claws and are even known to damage their feathers as a result.

The adult hoatzin also has some strange characteristics. It has an extraordinarily small head with a peculiar comb and it has eyelashes — a rarity among birds. In addition this bird has an enormous crop which is similar in function to that found in some herbivorous mammals such as cows. The bird's crop is so disproportionally large that when it is completely full of partially digested vegetable matter the bird is unable to perch.

Hoatzins have another rare characteristic. They exude a powerful musky odour which can attract predators. Despite the young birds' unique ability to escape, only about 50 per cent of them reach adulthood in one region of Venezuela. Their chief enemy is the black-capped capuchin which easily finds their nests because of their smell. The natives call it the 'stink bird' and most of them consider it unattractive meat, and generally leave it alone, although some tribes value the bird's flesh as a medicine. These people also collect hoatzin eggs, which they find delicious.

Hoatzins often live in colonies and have developed an unusual child care method based on voluntary service. A number of young hoatzins live without their own family and care for the other young. They assist in incubating the eggs, and in feeding the babies and are also always ready to defend them from predators. It is believed that 50 per cent of all hoatzin offspring are raised by so called 'helpers'.

It should be noted that although many characteristics seem to relate this interesting bird to prehistoric birds or even reptiles, studies indicate that their strange claws are a relatively recent development.

25.
Funny Creatures

Pocket Full of Miracles

The kangaroo is no doubt one of the most peculiar animals in the world. No wonder people were astonished when they saw it for the first time. Such was the experience of Francois Pelsaert, a Dutch sailor, who was shipwrecked onto Australian soil in 1629. As he was resting exhausted on the grass, he suddenly saw a number of spectacular creatures, which he described as having dog-like heads, very small, practically useless front legs and enormous tails and as hopping in a peculiar way.

Australian Aborigines are profoundly impressed by the animal's sexual behaviour. They claim that the kangaroo can copulate for up to two hours. This is probably why some Aboriginal males want their penises to look like those of the kangaroo. The kangaroo penis is two-headed (bifid) and the adult men in some Aboriginal societies mutilate their penises to look the same. As a result they are unable to direct the flow of their urine and have to squat to urinate. This peculiar practice is most common among the Aborigines living in central and northwestern Australia.

We owe the name of these strange animals to Captain Cook who asked the Aborigines what this peculiar animal was called. The natives simply answered 'kan ga ru' which, more or less, meant 'I do not understand'. Captain Cook assumed it to be the animal's name and it has been called 'kangaroo' ever since.

According to an Aboriginal myth, the first kangaroos used all four legs, but one day, when the animal was asleep, a huge fire broke out and surrounded it. The kangaroo managed to escape, but its front legs were so badly burnt that they became much shorter and since then he only uses his hind legs and hops.

Most Aboriginal tribes relish kangaroo meat and hunting kangaroos is a traditional activity. They often hide near water holes and spear the animals when they arrive to drink. Tracking kangaroo in the past was easy as they occurred in enormous numbers in herds of up to a thousand individuals.

Some Aboriginal tribes, however, highly respect the kangaroo and have the same name for it and humans. They do not hunt them and consider eating the meat an act of cannibalism. Some tribes believe the kangaroos are their ancestors. During the initiation ceremonies of a Kakadu tribe, an old man behaves like a kangaroo, imitating the noise they make and their hopping gait.

One of the many wonders of the kangaroo is no doubt its reproduction. Although it was known that the baby develops in the pouch, it was a mystery how the defenceless, blind infant, who is no bigger than a honeybee, found its way up to the pouch. It was believed that the mother placed her 'baby' in the pouch with her 'hands'. Finally, it was established that the minute infant needs no help. Using its forearms, it climbs up to the pouch. The largest species make the 15 cm or so long trip in just three minutes. Once inside the mother's pouch, the tiny youngster attaches itself firmly to a teat and starts drinking milk. It stays there for 6–9 months and during this period the tiny baby (of a large kangaroo species) will increase two thousand times its original size.

This is not the only surprise in her 'pocket full of miracles', as one author called it. Just two days after giving birth, the female can mate again. The kangaroo mother, however, holds the resulting embryo in an arrested state, because of the first baby in the pouch. When the first baby leaves the pouch (or dies) the development of the 'spare' embryo resumes and the second baby is born. From then on the kangaroo mother can regulate the chemical composition of the milk she secretes from her teats. She produces two kinds of milk for her two children — a low-fat milk for the new-born baby in the pouch and a high-fat milk for her older baby which feeds from outside the pouch for a whole year.

When the offspring is big enough it leaves the pouch, but still hides inside when danger threatens. One can often see a baby, which seems far too large, still trying to enjoy its 'old' home from time to time.

Although kangaroo mothers seem to be very devoted to their babies, to escape danger she can eject her young from the pouch to increase her mobility and chances of escaping unhurt. Eastern grey kangaroos (*Macropus giganteus*) are very fast, the record speed being 64 kmh. A big kangaroo can jump a fence 3 m high and can leap a distance of some 9 m. The kangaroo record in long jump is 12.8 m.

When endangered, these rather shy animals usually escape and rarely fight. If cornered by a human, however, a big kangaroo can be very dangerous. Bracing himself with his huge tail, he can kick with great force and a large kangaroo can disembowel a man and inflict fatal wounds with his sharp nails.

The biggest is the red kangaroo (*Megaleia rufa*) which is widely distributed throughout Australia. It is the world's largest marsupial, standing up to 2.1 m tall and weighing up to 90 kg. Despite its common name, only the male has red to greyish fur with white underneath. Females are bluish-grey and weigh not more than 27 kg.

In some regions kangaroos are so common that they have become the worst enemy of the farmers, who are forced to make thousand-kilometre-long, tall fences to keep kangaroos away from their crops.

Some kangaroos live very differently. There are two kangaroo species in Australia and a few in Papua New Guinea which spend most of their lives in trees. The Lumholtz's tree-kangaroo (*Dendrolagus lumholtzi*), lives in tropical forests near Cairns in Queensland. It is especially adapted to life in the tree canopy and is known to jump to the ground from a height of some 15 m when looking for water. It can leap from one tree to another with ease. The sight of an 80 cm long kangaroo perching in a treetop surprises naturalists and is a great tourist attraction.

Mammal with a Bill

The platypus (*Ornithorhyncus anatinus*), found only in Australia, is so different from other animals that when the prominent biologist, Dr George Shaw, was presented with a specimen by a sailor in London in 1799 he concluded it was a hoax. He was initially so convinced that it was a trick that to prove it he tried to cut off the duck-like bill with scissors — the scissor-marks can still be seen on this preserved skin in the Natural History Museum in London.

One should not be surprised by this scientist's attitude because what he saw was really unbelievable. The animal looked like a cross between a mole – as its body was covered with fur – and a duck, because of its duck-like bill. The Australian sailor who had brought the skin insisted that it was genuine and so Dr Shaw decided to consult another specialist. After a long and detailed examination and consultation, the two scientists finally concluded that it was a genuine animal. Dr Shaw wrote: 'Of all the mammalia yet known, this seems the most extraordinary in its conformation, exhibiting a perfect resemblance of the beak of a duck engrafted on the head of a quadruped'. In this way the scientific world learned, with astonishment, of this bizarre Australian animal.

The platypus is up to 60 cm long. It has fur and a large tail like that of an otter, webbed feet like aquatic birds and only one opening for ejecting faeces and urine, a feature of a reptile. It was later revealed that its young feed on milk, but it was hard to accept another fact — that this animal lays eggs. Travellers had learned more about the strange habit of this animal from local Aborigines, but this news caused a lot of controversy among biologists. A mammal laying eggs was just too far-fetched to be acceptable.

Finally, Dr W. Caldwell, from Cambridge, was sent to Australia to find platypus eggs. After a long search, he found some eggs, which had soft shells like those of lizards. With this 'discovery', the classification of this strange animal with the features of reptiles, birds and mammals, became almost impossible. Eventually, it was classified as a mammal. This peculiar semi-aquatic animal is known to have existed 50 million years ago and is considered to be a link between reptiles and mammals.

Although its bill is shaped like that of a duck, it is not as hard. It is covered with a soft, leathery skin and its function is entirely different from that of the birds. It is the platypus's sensory organ which it uses to detect other animals in the water. The animal swims with its eyes and ears closed — so cannot see or hear in water. As a result, its large bill is an essential organ for navigation. Its sensors are so sensitive that they can detect an electrical field as low as 500 million parts of volt. Any object moving in the water causes a slight change in the electrical field and can be detected by the animal's remarkable sensors.

Another surprising feature of the platypus is its poison spur, a strange feature again likening it to reptiles. Unlike snakes with their venomous fangs, platypuses have venomous spurs on each of their hind legs. Although both sexes have these spurs when young, as the female matures her 'sting' organs degenerate. In effect, only adult males have venomous spurs which they use in both defence and attack. The spurs are sharp enough to inflict a small wound, and the poison injected is strong enough to kill a small prey. Adult males are even known to use their venom against younger members of their own species.

The animal's sting is not lethal to men, but it is extemely painful and a number of people have had to be treated in hospital. According to one account from 1876: 'He / platypus / revenged himself by giving my friend a severe wound on the hand, one spur slightly tearing the palm, and the other the back of the hand, making a deep puncture between the knuckles / I think / the first and second fingers. The pain from this was so intense and almost paralysing . . . it was half an hour before he could stand without support. By that time the arm was swollen to the shoulder, and quite useless, and the pain in the hand very severe.'

The platypus is a very fragile and sensitive animal and it is said that it can be killed by a very loud noise. During the Second World War, this remarkable Australian animal was being taken, on board a ship, as a small gift to Winston Churchill. An enemy bomb exploded near the ship, and one hour later the small passenger was found dead. It was concluded that it could not withstand the loud noise of the explosion.

The platypus has interesting breeding habits. The female lays two or three eggs in a nest chamber 30 m long, winding burrow dug in the river bank. The incubation period is very short, lasting only about ten days and the blind and naked hatchlings are only some 2.5 m long. They can not find the teats like 'normal' mammals, and so the mother's milk is secreted from enlarged pores on her breast, which the babies lick from her fur.

Adult platypuses have insatiable appetites and spend most of their time searching for shrimps and worms, which are their main food. These strange animals need such a lot of food daily that sending live animals overseas can be difficult. When two platypuses were sent to the USA by plane the provisions for these 'passengers' alone included 10,000 earthworms, 500 mealworms and 550 shrimps.

The platypus, which is now much admired, was not, however, well treated in the past. Early European settlers soon realised that the animal's fur could be a good money maker. Once very common in Australian lakes and rivers, it soon became a rare animal. It was feared, at a certain stage, that it would become extinict unless protected by law.

Fortunately, the platypus is now a protected animal and great efforts are made to increase the indigenous animal population.

Several projects have also been established and the animals can be observed in artificially created sanctuaries. In 1991, a platypus was born in captivity in Australia, causing a great sensation, and biologists and conservationists were delighted. All the daily papers reported the event, with pictures of the famous baby. 'It took 15 people with a row boat, canoe and three big nets all night to catch the baby platypus – the first born in captivity for almost 50 years', according to one Australian newspaper.

Australia's 'Teddy Bear'

The koala (*Phascolarctos cinereus*) is a unique animal in many respects. It feeds almost exclusively on gum tree leaves and an adult can consume up to half a kilogram of fresh leaves every day. Chewing leaves seems to be its main activity as it spends most of its time curled up like a human baby high in the fork of a tree sleeping for some 18 hours a day.

Although there are over 600 species of eucalypts in Australia, koalas only eat a few eucalypt species completely, ignoring the rest. To cope with such a harsh diet, it has a specially developed caecum up to 2 m long. Its digestive tract houses special micro-organisms which digest the tough leaf material. Since many eucalypts contain highly toxic, volatile oil, it has been suggested that koalas become drunk and this is why they appear so sleepy. This is not true, as the animal has a fantastic detoxifying mechanism to counter the effects of the oil.

The koala was discovered in 1802 by F. Barrallier, but his description of this unique animal was not very elaborate, or enthusiastic, when it appeared in the Sydney gazette: 'It has a false belly . . . and its food consists solely of gum leaves in the choice of which it is excessively nice'. A full description was not published until a hundred years later. But when the first koala was brought to Europe it caused a great sensation, and not only among biologists.

It was at that time that the full details of the koala's strange breeding habits were revealed. As a marsupial, the koala is related to the kangaroo and other pouched animals. Like kangaroos, the youngster is minute when born, only some 2 cm in length. It also has the remarkable ability fo find its way to its mother's pouch. When the tiny, blind baby reaches the pouch it attaches itself to the nipple, which immediately expands and keeps the baby firmly attached in a 'drinking' position for some three months. It is carried in the pouch for a further 3–4 months and when the baby is sufficiently grown, its mother has to change its diet from milk to the tough gum leaves of the adults. To avoid too sudden a change, the young are weaned gradually and during this transition period, in addition to mother's milk, the baby is fed directly from the mother's anus. Instead of droppings, the mother starts to excrete a kind of partly digested leaf soup which contains sufficient nutrients to keep the baby in good health during the weaning period. Strangely, during this unusual feeding procedure, the mother's anus remains perfectly clean.

At the same time, the amount of milk given to the baby is gradually decreased and then completely stopped.

Some Australian Aborigines called this animal 'koolah' and probably the first European settlers adopted this name, changing it slightly. The Aborigines' attitude toward koalas differs depending upon the tribe. The Bidgara people of Queensland believe that it was a koala who brought rain and changed their barren desert into a most beautiful, forested land. Members of some tribes respected koalas highly and regarded them as wise counsellors, from whom they sought advice on the most important matters. Generally, in the past, however, koalas were hunted by the Aborigines for food.

The first European settlers initially admired these strange 'teddy bears', but they soon became more interested in their beautiful fur. Koalas are an easy prey and by using snares, poisons and trained dogs to detect them an enormous number were killed. To satisfy the overseas demand for their skins, eight million koalas were slaughtered between 1919 and 1922. The koala population decreased so dramatically that it was threatened with extinction. Fortunately, a law prohibiting koala hunting was passed in time and the famous animal was saved at the last moment.

Even now, the koala population is diminishing dramatically, not because of hunting but because of the intensive destruction of gum trees on which koalas feed. Fortunately, in the 1980s, a Koala Conservation Program was established and koalas can now live in peace in special sanctuaries. Financial aid for koala conservation projects has come from all over the world. In 1985, American Express International donated $200,000 'to help save the Australian we'd all hate to lose'.

King of Darkness

With their peculiar, membranous wings, and very odd expression, bats have always attracted people's attention. It was believed that a bat was half bird and half mammal, because it suckled its offspring and flew like a bird. There was even a myth that humans originated from these unusual creatures.

Bats used to be associated with darkness and death. The ancient Maya believed that the kingdom of darkness was ruled by a bat named Camazoth, god of death, who was said to decapitate his human victims.

This powerful bat-deity was worshipped by the ancient Maya and a city in Mexico was called 'Tzinacent Lan', meaning 'Bat City'.

The belief in the mysterious power of the bat-god has almost disappeared, but it still persists in some Indian cultures in Mexico. Large caves are believed to be the home of the bat-god, and pregnant women take offerings to these caves and pray for a successful delivery. For the Chinese the bat is a sign of good fortune and their symbols for a bat and for happiness are the same. Certain tribes in Central Africa believe that bats, involved in witchcraft, are associated with damage to crops. According to a Finnish belief, the human soul takes the form of a bat and escapes from the body during sleep. This is why, they explain, these animals cannot be seen during the day.

Biologists have always been fascinated by these creatures, the only mammals that can truly fly, and their unusual precision of flight in complete darkness. Despite the saying 'as blind as a bat', bats can see but they do not rely on vision alone. Their remarkable ability to detect a small fly in total darkness remained a mystery for centuries. Finally, an Italian naturalist, Lazaro Spallazani, deduced that bats must have special sensors which enable them to fly in darkness. He blinded some bats and observed that their flight and orientation did not change and so proved that they possessed something other than unusually acute vision. We now know that the bat has an echolocation system, which is a kind of radar. Although they appear to fly silently, they, in fact, send out ultrasounds which are beyond our hearing. These sounds are echoed by any object they hit and are picked up by the bat's highly sensitive ears. The clever bat judges the distance between itself and the object from the time it takes the sound to come back and then behaves accordingly. This remarkable skill enables bats to detect and avoid the smallest obstacles. In an experiment, a bat detected tiny wires only some 0.1 mm thick strung across a completely dark room. Incredibly, when the bat became acquainted with the configuration of the wires it closed down its radar and continued to fly accurately without its aid. This animal's system is a million times more sensitive than the sonar devices used by humans.

Some bat species live in large, dark caves where huge colonies nest year after year. Among the world's most populated are the Carlsbad Caverns in the USA, in which an incredible number of free-tailed bats (*Tadarida brasiliensis*) are known to have lived for centuries. Their long residence can be measured by the depth of guano deposits covering the cave floor, which is some 30 m. These large guano deposits attracted attention in the past. During the American Civil War, it was used for making gun-powder and the caves were guarded by a special military unit and treated as strategically important.

Most extraordinary perhaps, was the US plan to use the bats themselves. During the Second World War a special secret project was established in the USA. It was a kind of bat suicide squad, in which the bats were to carry incendiary bombs to destroy enemy industrial centres. Each bat carried a small bomb attached to its chest. The bats were packed with their bombs into wooden boxes and then parachuted over enemy territory. When released, the bats crawled into cracks in buildings, chewed off their harness and escaped before the bombs exploded. Preliminary, experimental tests proved successful as the parachuted bats sabotaged the places chosen for the practice. The project, however, was soon abandoned because the atom bomb was about to be completed.

It has not always been realised how beneficial bats are to humans. As most bat species feed on insects, they significantly reduce their numbers, including mosquitos and flies. A colony of 100,000 bats can consume up to 360 tonnes of insects annually. Some bats even catch fish, using their echolocation system. They are sensitive to the slightest disturbance or ripple, and, flying above the water surface, they suddenly drop their large feet into the water and grab fish swimming beneath.

Many bats are important seed distributors in tropical regions. After consuming fruits in forest regions they often 'plant' seeds with their faeces in deforested areas.

There are many fruit-eating bats in tropical Africa and, in some regions, they occur in such numbers that the trees from which they hang look as if they are laden with fruit. Native boys shoot these bats using slings. They are sold at the local markets and provide people with additional 'chicken' meat, which is cheap and readily available.

There are some 950 species of bat in the world, which vary greatly in appearance, size and habit. The largest, in terms of wing span, is the bat (*Pteropus vampyrus*), which is known as the large flying-fox because of its strange, dog-like head. Its wing span can be up to 1.65 m. This remarkable bat feeds on fruit and lives in New Guinea.

The smallest bat is the Kitti's hog-nosed bat (*Craseonycteris thonglongyai*) discovered in 1973 in Thailand. It has a wingspan of only 16 cm and weighs some 2 grams; it is in fact the world's smallest mammal.

Despite the fact that bats seem rather awkward to handle, some people, especially in tropical parts of Australia, admire these odd creatures and keep them as pets.

Mona Lisa of the Forest

In tropical forests of Central and South America lives a peculiar creature, called the sloth. It hangs upside down from tree branches and spends most of its life asleep as it is only active for about four hours a day. Its strategy to avoid predators is to remain unnoticed and so the sloth rarely moves. Sloths are in fact the slowest of all mammals.

The slowest among the seven sloth species is a three-toed sloth *Bradypus tridactylus.* It takes half a minute to move a leg from one position to another or to move 10 cm along a branch. The animal moves no faster on the ground than on the tree top, and it needs a minute to cover 2 m or 5 hours to cover one kilometre. In fact, the sloth cannot walk normally on the ground, but moves awkwardly by dragging its body forward with its claws.

The sloth's slow movement has always puzzled biologists and was accurately reflected by Archie Carr: 'It is a mockery of motion, an eerily mechanical, nerve-wracking slowness that contractile protoplasm was never meant to support. The cytoplasm of an amoeba streams faster than a sloth flees from a hungry boa constrictor. And besides being thus pointlessly unbearably slow in everything it undertakes, the sloth is hesitant and vacillating in undertaking anything'.

Plants grow on the animal's long fur, which is always wet because of the forest's humidity. Algae grow in abundance and provide the animal with an excellent camouflage. As the algae turn yellow during drought and green when it rains, the animal has an appropriate appearance during each season. From a distance, the animal looks like a bunch of dead leaves or a wasp's nest.

The green algae living on the fur of some sloths are attractive to certain insects. Some species of pyralic moth lay their eggs in the animal's hair where the hatched caterpillars feed on the algae.

The sloth performs all its essential life functions in its strange upside down position. It feeds in this position and also mates hanging upside down. When giving birth, the female also hangs from the tree branch and the baby could fall and be killed, but fortunately it has well developed claws with which it quickly catches onto its mother's long fur. It is a strange thing to observe a female cuddling her baby sitting 'comfortably' on her belly.

The sloth's toes have strong, hooked claws, which make its grip on tree branches incredibly firm. Vigorously shaking the branch on which a sloth is hanging will not dislodge it. Even if a sloth is shot dead while in its hanging position, it will not release its grip and will remain hanging until its body rots.

Almost everything about a sloth is slow or delayed and it is said to defecate only once a week. In order not to break its excellent camouflage by making a noise, it descends from the tree. Since it may spend its whole life in one tree, it also has one toilet which it carefully covers after each use to prevent the smell from indicating its presence to predators.

The animal is also said to urinate extremely rarely. According to some authors, it only urinates from the tree when it is raining which muffles the sound of the falling urine. The animal has a big bladder which can stretch enormously and hold a lot of urine for a long period of time.

The sloth is often called 'lazy man' by South Americans, because of its incredible slowness. But to the naturalists who admire this creature's unique qualities and its peculiar, 'half smiling' face, it is the 'Mona Lisa' of the forest.

Despite their efforts not to be detected, sloths are the prey of large snakes and harpy eagles. Nevertheless, they are difficult to dislodge from their branches and, as their fur is very thick, even eagles with their sharp talons cannot always kill them. Injuries which would certainly be mortal to other animal heal surprisingly quickly in this otherwise extremely slow creature.

Moreover, in sharp contrast to most other mammals, the sloth's body temperature varies significantly according to the temperature of their surroundings. The two-toed sloth has the lowest body temperature of all mammals, ranging between 23°C and 33°C.

Sloths date back some 50 to 60 million years and they were among the earliest mammals of South America. Ground sloths were as large as elephants and an extinct sloth, called Megatherium, was probably 6 m long and weighed up to 5 tonnes. The contemporary species of sloth vary in size. An adult three-toed sloth measures some 60 cm in length and usually weighs no more than 4.5 kg.

Floating Head

The ocean sunfish (*Mola mola*) has no distinct tail and looks as if the rear half of its body has been cut off just behind the dorsal and anal fins. What remains looks like a gigantic floating head and fishermen call this bizarre creature a 'headfish'. Even the arrangement of its internal organs is unusual as, for instance, the kidneys are just behind its brain. These floating heads can be enormous — the largest specimen, caught in waters off Sydney, was 4 m long and weighed about two tonnes. It is the heaviest bony fish in the world and also the most fertile. An average sized specimen, some 1.3 m in length, was found to contain as many as 300 million eggs — it is not difficult to imagine how many more eggs a giant ocean sunfish can produce.

The eggs are extremely small, only about 1.3 mm in diameter, and the freshly hatched larvae measure only about 3 mm. It has been estimated that its adult weight is 60 million times that of the larvae. The larvae are so different from the adult form that in the past they were mistakenly classified as a different species. They have distinct tails which disappear during its development when their bodies become covered with elongated spines. When the larva is about 1 cm long the long spines disappear and a new fringe appears. The larva's body then gradually changes until it finally acquires the peculiar shape of the adult fish.

Ocean sunfish swim very slowly and cannot escape predators. Instead, they have another form of protection. Their leathery skin may be over 5 cm thick and is said to be so tough that it can even withstand rifle bullets. With such a cover the fish is reasonably safe from predators. On the other hand its skeleton is so soft that the bones can be cut easily with an ordinary kitchen knife.

Ocean sunfish have an insatiable appetite and grow very quickly

They particularly like jellyfish and are specialists in dealing with this prey. They first bite off the jellyfish's tentacles and then start devouring its body. This food is not very nutritious, however, because jellyfish are almost 98 per cent water and, to satisfy its needs, the sunfish spends almost all its time feeding.

These slow-swimming large fish are an easy catch. Their flesh, however, is leathery and considered inedible, but the Japanese catch them for their livers, which they consider delicious. These livers are very rich in oil and, in the past, whale hunters used to catch 'floating heads' if they were unable to find a whale. The liver, and the adjacent tissues were cut off and placed in barrels in the sun where they gradually broke down and eventually clear oil would collect at the bottom of the container. The ocean sunfish oil was not only used for softening leather, but also as a cure for rheumatism.

26.
Giants of the World

Living Island

A group of sailors were once hopelessly drifting in their rescue boat after their ship had sunk, when they suddenly noticed, with great relief, a small, barren island. The sailors quickly landed, but as they began breaking up the boat to make a fire, the 'island' started to move. They quickly realised that they had disturbed the tranquillity of an enormous whale, the gigantic mammal of the ocean.

Although the above story is a typical seafarers' fable, it nevertheless shows how enormous a whale can be. The blue whale (*Balaeroptera musculus*) is the largest creature on earth, measuring up to 30 m in length and weighing up to 150 tonnes, which is equivalent to the weight of the total population of a small town of some 2,000 inhabitants. The biggest dinosaur weighed 'only' about 75 tonnes.

All the internal organs of the blue whale are of an incredible size. Imagine a heart the size of a small motorcar, pumping some 15,000 litres of blood into huge arteries which are wide enough for a small baby to crawl through. Its remarkable tongue can weigh up to 4 tonnes and the male's penis can be a record 2.4 m long.

The world's largest testicles belong to the right whale (*Balaena glacialis*), which can be up to 2.7 m long and weigh up to half a tonne. This whale species is only some 15 m in length.

The new born baby blue whale also makes a world record, since it can be some 8 m long and weigh up to 3 tonnes. Feeding such a huge baby is a demanding task as it sucks some 130 gallons of milk daily. The baby grows extremely fast and its nursing mother can lose up to 50 tonnes of her body weight in the process.

Despite its enormous mouth, the blue whale cannot swallow anything bigger than small fish or krill, but it consumes up to 4 tonnes of such food daily. To obtain this quantity of food it filters some 5,000 litres of water at every gulp, but it only feeds for 4 to 5 months annually.

These enormous animals communicate with each other by emitting low-frequency pulses, which are the loudest sounds produced by any living creature. These pulses can be detected from a distance of over 850 km.

There are many species of whales, but the one which was most extensively hunted in the past was the sperm whale (*Physeter catodon*) the largest of the toothed whales. This whale has the heaviest of all brains. It can weigh over 9 kg, which is six times heavier than an average human brain. Its head is about one-third of its total length and contains an enormous amount of pure oil — up to 2,300 litres. Early whalers incorrectly thought that the oil was the creature's sperm, hence its common name. While the sperm whale can weigh some 70 tonnes, the organ containing the oil can weigh as much as 16 tonnes. This oil, called spermaceti, is of great importance to the animal since it enables it to remain motionless in water without using much energy. By changing the oil density, the animal can change its buoyancy and dive to a great depth. It can spend up to one hour under water and it has been estimated that the sperm whale usually dives to a depth of up to 1 km, although the record dive was 3 km.

The sperm whale's lower jaw is equipped with powerful, conical teeth, which can be up to 20 cm long, the largest mandibular teeth in the world. With such a well-armed jaw the sperm whale can catch and eat large squid and fish as big as some sharks. Even with damaged jaws they have been known to capture their prey. In addition, it has been suggested that this animal can immobilise smaller prey, such as fish, with a burst of high-intensity sound.

These enormous sea mammals were held in great respect by many nations and in some regions they have even been worshipped — from the Ainu aboriginal people of Japan to the Indians of Peru. Among some groups of Eskimos, if a hunter killed a white whale he had to observe certain taboos, similar to those observed by mourners. Certain rituals were also customary before hunting whales. Fasting, abstaining from liquor and even sexual abstinence had to be observed. To purify the soul before the dangerous fight, each hunter had to confess his sins to the others and, if the sins revealed were too serious, he was excluded from participating in the expedition.

If the bodies of Alaskan Koniaks who lost their lives during a hunting expedition were recovered, they were preserved in secluded caves. In addition, just before an expedition to hunt whales, the Koniaks placed the bodies of former whalers in a stream and the water was drunk, in the belief that it would bring them luck. In some regions, the body of a hunter killed by a whale was cut into small pieces and dried and then worn as amulets to bring success in whale hunting.

Even after a successful whale catch, special taboos were held among hunters in some regions. The inhabitants of a certain region of the Bering Sea islands believed that iron knives should not be used for cutting whale's flesh as it would offend the whale's soul. Even making loud noises while dismembering the animal's body was prohibited in order not to frighten the animal's spirit.

Whale hunting was a big business and the demand for whale oil, at the beginning of the nineteenth century, was enormous. The most important man on the whaling ship, after the captain, was no doubt the harpooner, as the success of the whole expedition often depended on his skill. The harpoon had to be thrown up to 9 m to hit the whale. Once hit, the animal would immediately dive and the line attached to the harpoon would run out with astonishing speed. The slightest swing of the rope could easily take off a man's arm or leg and harpooners were sometimes tossed overboard by the running line. It could take a day or longer for the huge animal to die after being hit and enormous effort and skill were required from the hunting crew before the huge animal would give up.

The slaughter of whales used to be enormous. Some 70,000 whales were killed around the world in 1962 alone. The most vulnerable was the giant blue whale and, at one time, its total world population was reduced to only some 6,000, from the original 200,000. Fortunately, the total prohibition of hunting the blue whale in the southern hemisphere prevented a further decrease in number of these remarkable giants of the sea.

Most people show great admiration for whales and whenever there is a threat to their lives strong efforts are made to help them. In 1985 some 3,000 beluga whales were trapped in the ice off the Bering Sea and Soviet icebreakers arrived to clear them a way to freedom. The crew used loud speakers, playing military and jazz music as a lure, but, strangely, the trapped whales were not tempted to use the escape route. Then, when classical music, including Beethoven, was played the animals started to respond and swam out of their 'prison'.

On another occasion, when three young whales got stuck under the ice in Alaska, the story of their fate was televised around the world. To enable the trapped whales to breathe, volunteers cut holes in the ice. Finally a Soviet icebreaker arrived and after some two weeks of hard work, two of the three whales swam to freedom.

Strange Cavalry

With its massive body, peculiar trunk and enormous tusks, the African elephant (*Laxodonta africana*) is among the most fantastic animals in the world. It is not only the largest land animal, but also the heaviest, as it attains the incredible weight of six tonnes. Its enormous tusks can reach 3 m in length and the heaviest single tusk ever recorded weighed 117 kg.

No wonder such powerful animals were used in warfare. In many regions, especially Asia, tamed Indian elephants (*Elephas maximus*), were commonly seen on the battlefields. The elephant 'cavalry' used to be so essential in some regions that the power of a ruling monarch was measured by the number of elephants he had in his army.

War elephants were specially prepared for their role and their huge heads and sides were covered with protective plates of thick metal. Their trunks were equipped with swords and poisoned daggers were tied to their enormous tusks. Such living tanks had an enormous physical and psychological effect on the enemy troops. In some Asian countries up to 2,000 elephants were used in a single battle and their power was tremendous.

According to a passage from Mahummud Casim Ferishta's *History of Hindustan*, written in 1768, the effect of war elephants was terrific: 'The terrible animals trod the Rajaputs like grasshoppers under their feet, or winding them in their powerful trunks, tossed them aloft in the air, or dashed them against the walls and pavements.'

When Hannibal took his army through the Alps and defeated the Romans in 218 BC, his 37 elephants were crucial. The Roman soldiers had not seen such 'monsters' before and fled in terror. When Caesar invaded the Britons in 54 BC, he had one huge elephant, protected with heavy iron armour and bearing a tower containing several archers and slingmen. The Britons were terrified by this elephant and their horses bolted at the sight of such an enormous beast. The Romans were able to cross the river Thames and thus just one elephant won the whole battle.

The ancient Assyrians were aware of the importance of the elephants in war, but did not possess any. When their queen, Semiramis, invaded India in the ninth century BC she cleverly had dummy elephants made to frighten her enemies. These dummies were straw-filled skins mounted on camels to imitate huge elephants, and had a tremendous psychological effect on the troops.

Even during World War II, the Japanese had a few squadrons of elephants which they utilised in their campaign in the Burmese jungle. According to them, soldiers on foot could make no significant progress in this warfare.

In the northern regions of India and in Thailand, elephants are trained for various tasks and are especially useful for transporting timber. Thailand is famous for its elephant training. It takes a long time and starts when the elephant is about six years old, when he is given his master who, surprisingly, is a small boy of the same age. The boy will be the huge animal's only master, trainer and devoted friend. They live and work together in harmony for many years.

The elephants are mainly used in forestry work where they are especially useful for felling trees and pushing and hauling logs with their powerful tusks. Indian elephants have been used on a large scale. For example, 20,000 tamed elephants were engaged during the construction of the famous temple, Angkor Wat, in Cambodia.

The Indian elephant is frequently used for carrying people, especially for hunting, during national festivals or just for carrying tourists in Indian national parks. But, in the past, Indian elephants were also trained to kill people. In fact elephants were often the state excecutioners of criminals. The elephant was first trained using dummies and, when commanded, crushed the victim with its powerful fore-foot. A rather macabre execution by an otherwise very friendly animal.

African elephants, which are difficult to train, are famous for their enormous tusks, which have been the source of ivory since ancient times. Indian elephant tusks are much smaller and found only in the males as the females' are only very short and do not extend outside their mouths. Sometimes elephants have an abnormal number of tusks and can carry up to seven small tusks.

These intelligent animals seem to know that they are hunted for their valuable tusks. According to one author, when elephants find tusks on the ground they carry them away and bury them. In one case, animals were observed smashing the tusks of their mates against rocks. During our stay in East Africa, we were told by one of the national park officers that when the feet and ears of an elephant killed by poachers were stored in a park shelter, elephants came and took the mate's 'remains' and carried them off to bury them in a remote part of the savannah.

Elephants are also known to show unusual compassion towards their suffering mates and help each other when necessary.

For example, some were observed to 'organise' assistance for a wounded companion. Two elephants got on either side of a weak animal and lifted him up to carry him to another place.

The most peculiar part of the elephant's body is its remarkable trunk which acts as a fifth limb. It is so sensitive that the animal can smell water from a distance of up to 4 km, and it is such an excellent 'hand' that it can even pick up a tiny leaf from the ground. It is also so strong that it can lift a heavy log. The trunk has some 40,000 muscles and is remarkably flexible. Its main function is to convey food and water to the mouth, and the animal has no problem reaching high branches when feeding itself.

Thousands of these magnificent animals have been slaughtered in the last few decades because of the great demand for ivory. Fortunately, the ivory trade has been banned in most countries and elephants are protected by law. Poachers still kill many, however, and, despite prosecution, ivory is still smuggled from Africa. Probably the best demonstration against the illegal ivory trade was given by the Kenyan President, Arap Moi. He personally set fire to a huge pile of ivory captured from poachers, which would have been worth millions of dollars if sold on the black market.

Legendary Man-Eater

The giant clam *(Tridacna gigas)* of the Pacific Ocean coral reefs is an enormous bivalve which can be 1.3 m long and weigh up to 330 kg. Terrifying stories have been told about them by pearl fishers who claim that they are very aggressive and attack unwary swimmers and divers. Men are said to have lost legs after stepping on hidden, open clams which clasped them tightly in their steeltrap jaws. Its grip was supposed to be so strong that a victim would be unable to open it and so would be trapped and drown. The people of the Pacific Islands still believe these stories and they call the clam a 'man-eater'.

While it is true that the giant clam does close its valves firmly, it does it quite slowly and a man stepping into one would always have enough time to get out. The stories of men being trapped and losing their legs cannot be true.

To test the claim about the giant clam's terrific grip, an Australian biologist made a plaster model of a human leg and inserted it between the valves of a very large clam. The reaction was slow, but the grip was so firm that the 'leg' could not be released until the clam's muscles were cut deeply. The 'leg' had deep cuts on it from the rims of the enclosing valves.

The world's largest giant clam shell can be seen in Paris. In front of St Sulpice cathedral a giant clam shell is used as a font for holy water. In the past, Indian maharajas used big clam shells as hipbaths, and the natives of the Pacific Islands use the shells as bathtubs for their children. There used to be a custom that, when a chief's son was born, the child had to be first washed in the sea and then in a giant clam shell.

Occasionally, giant clams can bring great fortune. In 1934 a diver found a very big pearl inside a clam he had caught in the Philippines, off the island of Palawan. Although the pearl was enormous, (24 x 14 cm) and weighed 6.4 kg, he did not realise its value and sold it for just a sack of rice. When an American collector learned about this find, he rushed to the Philippines, hoping to buy the pearl. Despite the huge sum of money offered for it, the pearl's owner would not sell. Soon after, the owner's son became seriously ill and the man called the wealthy American for help. Immediate assistance was provided and the son recovered. His grateful father then offered the giant pearl as a gift to his benefactor. He probably did not realise how precious his pearl was. It is the world's largest pearl and, despite its irregular shape, its value is enormous. Called the 'Pearl of Laotze', it has changed owners often and is now the property of a Los Angeles jeweller. According to a report published in 1990, it was valued at US $55.7 million.

Between its two valves the clam has a substantial amount of muscle and fleshy tissue and a big clam can provide several people with excellent meat. As Captain Cook once said 'more than two men could eat' the flesh from one clam, and the one he saw was rather small.

Gathering giant clams for food used to be so common in the Pacific region that there was a danger of their extinction. To prevent this, their collection has now been prohibited in most countries. In recent years, however, Taiwanese poachers alone are said to have gathered a million giant clams.

Some fishermen are currently establishing their own clam farms.

In Papua New Guinea people collect giant clams when they are small and place them in special areas near their villages. The clams grow very slowly and only increase some 5 cm a year. To have an edible clam they have to wait for about ten years, but to become a 100 cm giant, clams need to grow for at least twenty years.

Giant clams mainly feed on plankton, but they also have another source of food. They establish a kind of a garden on the soft parts of their bodies where they cultivate tiny unicellular algae (Zooxanthellae). These microscopic algae live inside the soft mantle of the clam and are constantly watered by the host. To enable their green partners to photosynthesise, the clams usually choose shallow sea waters so that the algae have plenty of sunlight. The clam never 'eats' all the algae and provides them with excellent conditions for growth and development.

There is an interesting river-dwelling clam which needs the 'help' of another creature for the development of its offspring. Its larvae must be transferred to the gills of a large fish in order to develop. To do this, the 'cunning' mollusc uses a strange trick. It imitates a fish. It has an extension on its back which waves like a fish tail, and it even has a false eye on its body. When a large fish approaches it, and opens its mouth to snap up the small 'fish', the disguised mollusc immediately fires thousands of her larvae into its mouth where they attach themselves to its gills. The mollusc's larvae stay there for some time and continue their development, until they eventually leave the fish and develop further in the water.

River Giant

The hippopotamus (*Hippopotamus amphibius*) is the most dangerous water creature in Africa. It has been estimated that hippos are responsible for more deaths on this continent than all other water animals combined. His jaws are powerful enough to kill the largest crocodile and would defeat a huge shark if attacked, as for example in the mouth of the Santa Lucia river in South Africa. A hippo, when cornered, can bite a man in two. If a farmer misses his shot, a hippo may escape initially, but later come back to take revenge, flattening the farmer's crops and destroying his hut. Hippopotami are the heaviest creatures on land after the elephant and can weigh up to four tonnes.

Hunters know how difficult it is to kill this huge beast. Before hunting hippos, members of the Thonga tribe used to be allowed to have sexual relations with their own daughters. They believed that this odd practice would give them the necessary magic power to conquer the dangerous beast.

Although the hippo is perfectly adapted to live in water, it is also comfortable ashore and, despite its apparent clumsiness can run faster than man.

Hippos generally only wander for a few miles to search for food, usually at night. One hippo, however, called Huberta, became a famous traveller in South Africa. He left Zululand in November 1928 and arrived at the Keiskama river in April 1931, covering some 1900 km. The inquisitive hippo visited homes, schools, and shops, peeping into windows and creating a great sensation wherever he arrived. He seemed to appear, as if from nowhere, and then to hide in water pools. This harmless hippo became so popular that he was protected by law and became a sort of national hero. Unfortunately, the famous traveller was killed by a man who took him for an ordinary hippo damaging his crops.

This extraordinary beast was admired in ancient times. The Egyptian goddess of childbirth was depicted as a pregnant hippo and, to ensure a lucky delivery, pregnant women used to wear amulets in her shape.

The Romans imported these animals to show in their arenas, and used the hippos' tough skin, which is up to 5 cm thick to make buckles and helmets for the Roman soldiers. In Africa, the locals often make walking sticks from long pieces of hippo skin. To make it straight, they hang up the fresh skin with a heavy stone attached to one end. When it is dry it is rounded and polished and the finished walking-stick looks like amber. Occasionally, the tusks of huge hippos, which are similar to ivory, have been used for making artificial teeth.

Hippos are social animals and live in very large groups of up to 100 individuals. Their groups consist of many cows, calves and juveniles of both sexes, but very few bulls. Unusually, females, not males, lead hippo groups and are responsible for their territory. The unwanted bulls live apart on the periphery of the group.

The male must always show his respect towards the female. When a bull enters the crêche and the female rises to her feet, he must obediently lie down or he would be forcefully driven away. Hippo females invented the babysitting custom long before humans discovered the idea. The youngsters are under strict care and, if a mother has to leave the crêche, another female will 'nurse' her baby and any other female's babies. Hippos are excellent mothers and show great devotion towards their offspring. To escape the potential attack of deadly crocodiles the baby often rides on its mother's huge back.

The calves are occasionally born under water. It is probably one of the ugliest of infants since it has a disproportionately enormous head and is fat. The baby often suckles underwater, taking its food for a minute and then coming up to the surface to breathe for a while before diving back again. The hippo mother shows a very human-like attitude towards her baby. When it misbehaves she scolds it and may even gently 'smack' its head if the baby is naughty.

Although they are often said to 'sweat blood', hippos do not possess sweat glands and the so called 'sweat' is a sticky, pink secretion from glands under their skin.

Hippos have an incredible appetite and the amount of vegetable matter they consume is enormous. An adult hippo can eat some 70 kg of fodder daily. His stomach measures up to 3 m in length and has a capacity of up to 190 litres of liquid.

Despite the damage caused by these huge animals on fields and crops, they can be at least of some benefit to people. By eating enormous amounts of vegetable matter, they create channels through the masses of water plants such as papiri, and in this way they prevent blockages forming and improve navigation in some rivers, ponds and lakes.

It should be mentioned that there is a second species of hippopotamus known as the pigmy hippo (*Choeropsis liberiensis*) which is only about a quarter the size of a big hippo. It lives in forest streams of West Africa.

Tree Chicken

The first visitors to the Galapagos Islands were astonished by the enormous numbers of giant lizards which literally covered the rocky beaches. Darwin, who explored these islands, wrote: 'I cannot give a more forcible proof of their numbers than by stating that when we were left at James Island, we could but for some time find a spot free from their burrows on which to pitch our single tent.' The creatures Darwin saw were marine iguanas (*Amblyrhynchus cristatus*), large lizards, up to 1.5 m long. Usually black or dark grey, they have a distinct crest that runs from their neck to their tail, blunt snouts and heavy bodies and really can be likened to a dragon in a fairy story. Even hungry sailors who landed on the island were reluctant to eat these giant lizards.

Marine iguanas are unique in that they are the only true marine lizard, since they are dependent on the sea for their survival. They only feed on seaweed and algae and are excellent divers diving up to 9 m in their search for food. When they appear in large numbers they almost blacken the water surface, returning to the same place on the shore after each dive. They are well adapted to the intense heat as their skin is water tight and they can regulate their body temperature to prevent overheating. While they are basking, red crabs often climb their bodies, pulling out ticks from their rough skin. Marine lizards spend most of their time on land and only rarely go into the water because of sharks.

The largest iguanas, however, live in tropical regions of America. Known as the common or green iguana (*Iguana iguana*), they may be as much as 2 m in length. Despite their bulky bodies, they are excellent climbers and, when threatened, can climb some 15 m up to the tree canopy. Even from such a height they can drop down in an emergency and escape unhurt.

Iguana meat is considered a delicacy and the animals are extensively hunted in South America. This tree-dwelling iguana is such a popular source of food that it is often called the 'chicken of the tree'.

The meat of various iguana species is considered to improve men's sexual performance, so it is used to cure impotence, as well as being eaten.

Various techniques, including snares, are used to catch tree iguanas. Since iguanas' main enemies apart from man, are hawks, hunters often imitate their cries which frighten the iguanas and they become motionless long enough for the hunters to catch them. Some Indians hunt iguanas with bows and arrows and use trained dogs to find them in the treetops.

In some regions, especially in northern Colombia, people are not interested in the iguanas' flesh, only their eggs, which they consider delicious. They like eggs which have not been laid yet, so they ignore the males and hunt mature females. Once they catch one they cut her open with a sharp knife and pick out the eggs from inside her body. After this cruel operation, the animal is allowed to go free, but the severely wounded animal usually has little chance of surviving.

Some iguanas have a bad reputation. For example, the spiny-tailed iguana, native to Mexico, often lives in old cemeteries. Although they only innocently perch among the old tombstones, some local people believe these 'horrible' creatures dig up and eat the buried corpses.

Certain iguanas in the south-western desert regions of the USA are called Chuckwallas by the local Indians. These iguanas (*Sauromalus* species) are fat and only some 30 cm long and are hunted by the Indians for food. Chuckwallas hide in rock crevices, wedging themselves in by inflating their lungs. To get them out, the Indians puncture their lungs with long pointed sticks.

On Petite Gonaive Island, off the coast of Haiti, the locals catch large iguanas by pouring rum into hollows in rocks along the shore. The iguanas are said to like the strong alcoholic drink very much and drink it until they are completely drunk. They are apparently very dangerous when sober.

Barking Worms

Most people are familiar with normal earthworms, which are especially appreciated by fishermen who use them as bait. There are some earthworms in Australia, however, which, at first sight, look more like snakes than earthworms as many of them are over 3 m long. The largest ever measured was 3.9 m long and only some 2.5 cm thick when fully extended and when contracted became as thick as a man's wrist.

The first Australian settlers were convinced that these long creatures were snakes, and could not be gigantic earthworms. Even when a detailed, scientific description and photographs of these strange earthworms were sent to the USA, the biologists were sceptical and reluctant to accept the facts. Only when a live specimen reached the USA was the existence of the monstrous earthworm officially accepted. These earthworms caused a great sensation and when they were exhibited for the first time in Melbourne, thousands of Australians came to see them.

The largest are the earthworms of the *Megascolides australis* species, which occur in South Gippsland in Victoria. They live in burrows which can easily be found by the kind of small 'volcanos' at their entrances. The giant earthworms do come to the surface, but, if disturbed, they quickly retreat underground. As they do so, they make a strange kind of underground groaning which can be heard by passers-by and can be frightening to those unfamiliar with these worms. As a result, these worms are often called 'barking worms'.

Another species of gigantic earthworm, *Didymogaster sylvaticus*, which is only some 1.5 m long, lives in the coastal forests of New South Wales. It is very similar in appearance to the gigantic earthworm of Victoria, but behaves differently. When disturbed, this worm shoots out a liquid to defend itself from a predator. The shot of liquid can travel up to about 1.3 m and may distract some predators, but it does not harm humans. This trick deters birds, who usually avoid them, probably taking them for snakes. Only the famous kookaburra is unafraid of these giant, squirting worms and shows great skill in handling them. It takes a worm in its beak and gradually swallows it in portions, while the rest of the worm dangles from its beak. Giant earthworms have a strong creosote-like odour. It is believed that it comes from the fragrant oils in their diet, which is mainly composed of leaves and roots of various forest plants.

Although giant earthworms are so large, their structure is very similar to that of ordinary earthworms. They also show the same remarkable capacity to regenerate. If a gigantic earthworm is cut into a dozen pieces, a new earthworm will develop from each section.

The world's first museum devoted to this worm has been established near Bass in Victoria and is, itself, in the shape of a giant worm. Giant worms can be seen live in large tanks and the owners of this peculiar display hope to attract both Australian and foreign tourists to see 'the wonder from down under'.

The greatest enemies of 'normal' earthworms are moles, which often catch more than they can eat. Instead, they bite off the head of each captured earthworm, thereby immobilising them. The mole can then store the live worms for a short time, until he wants his next meal. If the mole 'forgets' about its larder, the earthworms regenerate their front ends and escape.

Earthworms were once used as medicine. A writer in ancient Persia reported: 'Baked and eaten with bread, they reduce the size of stones in the bladder. In difficult labour they bring on delivery immediately. Their ashes applied to the head with oil of roses make the hair grow.'

The largest specimen of earthworm ever recorded was the *Microchaetus rappi* of South Africa which measured 6.7 m.

Chinese Dragon

The world's largest lizard remained unknown to the western world for several centuries. European sailors' stories of spectacular monsters living on certain islands of the Dutch East Indies were ridiculed and it was not until 1912, when a Dutchman, P. A. Ouwens, brought an entire giant lizard skin to Europe, that its existence was accepted.

This spectacular lizard, the largest specimen of which measured 3.1 m, and weighed 166 kg, became known as the komodo dragon (*Varanus komodoensis*) or komodo monitor. With its enormous head and forked tongue flickering like a flame, this animal was likened to the fire-breathing dragon of oriental mythology. It was probably this animal which inspired the famous paper dragons in Chinese ceremonies.

It was commonly believed in ancient China that this dragon was the lord of water and was responsible for bringing rain. Even nowadays, a belief persists that if someone is going to build a house, he must first check that a dragon does not sleep on the proposed site. A priest is usually invited to examine the site and it is only accepted by the owner when the priest reports that there is no dragon present.

Some people liken the world's largest lizard's ferocity to that of the lion. It can bring down an adult water buffalo, weighing some 700 kg, using, however, a special technique. The giant lizard creeps up to a buffalo unnoticed and bites and severs its hind leg tendons so that it falls to the ground, unable to stand. The lizard then withdraws. Since the buffalo is unable to move, it cannot feed itself and after some time it starves to death. The monitor patiently waits nearby and eventually devours the rotting carcass.

When dealing with smaller prey, such as a domestic pig, the ferocious lizard brings it down and immediately starts eating it while it is still alive. Those who have seen it in action say it is a horrible spectacle.

Their favourite prey are new-born, wild horses. The lizard carefully observes a pregnant mare's behaviour and waits for the birth. When the foal is just about to drop to the ground, the powerful lizard snatches and devours it. The lizard seems to realise that the mare is too exhausted after delivery to defend her baby.

Komodo dragons are also dangerous to humans. Scientists have had to escape from their tents when these lizards attacked them and forest workers are occasionally attacked from behind by large monitors. Although they usually survive, their clothes are torn to shreds by the giant lizards and there have been cases of lethal woundings. When hungry, the large monitors are known to attack and devour their weaker mates, but when plenty of food is available, these lizards eat so much that they are unable to move for several days.

Adult komodo dragons have a very acute sense of smell and can detect rotten meat from a distance of up to 8 km. They also have good vision and can spot their prey from a significant distance by 'standing' and raising their huge heads to scan the surroundings.

The remarkable monitor has been protected since 1925. They are hated by the local inhabitants, because they steal pigs and goats from their farms and even dogs cannot prevent a hungry monitor swallowing a goat in just a few gulps. Its mouth has such an enormous capacity to expand that it can swallow a small pig.

There are also many impressive lizards in Australia. The largest is the perenty (*Varanus giganteus*), a formidable lizard which can be up to 2.4 m long. It is the second largest lizard in the world. Although its powerful tail is strong enough to throw a man off balance, it rarely attacks humans.

Another large lizard is the lace monitor (*Varanus varius*), which was once common on the outskirts of Sydney and can be found throughout the eastern parts of the continent. This lizard may be up to 2 m long and is the largest tree-dwelling lizard. Its instinct to climb is so strong that if there is no tree around when it is disturbed, it may climb up any object, including a horse or a man.

These large lizards are common in some regions of Australia and are hunted by Aborigines, who regard their meat as a delicacy and also eat their eggs.

Sea Monster

Giant squids (class Cephalopoda) have frightened sailors since ancient times. Their huge, torpedo-shaped bodies and enormous, forward-facing arms are extraordinary — no wonder so many incredible stories have been told of giant squids overpowering passing boats and devouring entire crews. For people with imagination, this strange animal of the deep became a mythical sea monster. In Norway, it became the famous 'kraken' — a most ferocious beast, which was said to rise up and attack sailors. Herman Melville, in his novel *Moby Dick*, described a giant squid as 'a vast, pulpy mass, furlongs in length and breadth, of a glancing cream colour, floating on the water, with innumerable long arms radiating from its centre, and curling and twisting like a nest of anacondas, as if blindly to catch any helpless object within reach'.

Such descriptions are not really accurate since the squid only has ten arms of which two are greatly elongated, and although some have reportedly landed on the decks of sailing ships, they are rarely dangerous to humans. Only one ferocious attack of a giant squid was verified. In 1941 a British ship was sunk and a few survivors were stranded, sharing a small rescue boat. When night came, a huge squid suddenly appeared and trapped one of the sailors in its monstrous arms. It pulled him below the water and the screaming victim was never seen again. Lieutenant Cox, who witnessed this accident, was also briefly entangled by this monster, but managed to escape from the embrace. Circular scars remained on his leg for many years after this tragedy.

Among the many species of squid, the largest is the Atlantic giant squid (*Architeuthis dux*) which is in fact the world's largest invertebrate. A giant specimen captured in 1878 was over 16 m in total length and weighed some 2 tonnes. These huge squids have incredibly large eyes. One specimen's eye was 40 cm across, which is the largest eye of any living or extinct creature. In comparison, the eye of a huge whale is only about 12 cm in diameter. It is not only this record which puzzles naturalists, as the eyes of some species show a great disparity in size. For example, in the sail squid (*Histioteuthis* species) one eye is five times bigger than the other, a phenomenon which cannot be explained. It is also strange that, despite the fact that squids are so remote from humans in their evolution, their eyes are even better, as they can distinguish much finer detail than humans.

Although people were usually frightened of the huge squids, some much smaller squids are believed to be more dangerous. A species of squid (*Ommastrephes gigas*), found in coastal waters of Peru, is regarded by local fishermen to be a real monster. It is only some 3.7 m in total length, but is greatly feared and it is said that if a man falls overboard he would 'be torn to pieces in less than half a minute'. Although nobody has yet been captured by this squid, its bad reputation stems from the fact that it can bite through the fishermen's tough wire-like nets.

Squid are famous for their interesting method of defence. When a small squid is under attack it does not immediately try to escape, even though it is a good swimmer. Instead it discharges a blob of black ink from the mantle cavity in front of its mouth. The shape of the ink blob which suddenly appears is very similar to that of the squid itself and distracts the predator. This gives the squid sufficient time to dart quickly away. Moreover, once it has discharged the ink, the squid becomes very pale, almost transparent and the predator usually cannot see it escape. The squid has a special gland which produces and stores the ink and can discharge it at will.

The 'ink' produced by squid was commonly used in the past as a dye and was regarded as an excellent and durable material. Its durability is incredible. A fossil squid was once found which still contained some ink in its body. The ink was still of good painting quality, after millions of years. In Spain there is a famous seafood dish made of squids which is called 'calamaros en su tinto' which means "the squid in its ink".

Squid is among the most important sea foods. In Japan some half a million tonnes of squid (mainly of the *Loligo* species) are caught every year for consumption. According to some biologists, squid may save the expanding human population from starvation in the future as it is a source of valuable proteins.

Symbol of Longevity

Turtles and tortoises are ancient animals which have survived for over 200 million years, and one of the secrets of their survival lies in their incredible armour which provides them with excellent protection against predators. Their shells cover both their backs and their bellies. The giant turtles' massive shells have always fascinated people and there was a belief that gigantic turtles carried entire islands on their backs.

There are many myths and legends surrounding these unusual reptiles. According to one tale, when Shih Huang Ti, the emperor of China (221–207 BC), was constructing the great wall, a giant tortoise offered to support the wall on its back, and, as a result, the Chinese called it the 'tortoise wall'. In Chinese mythology, the tortoise was one of the four creatures who created the earth. This animal was also linked in China with death and burial as it was believed to be the messenger from the after-life, and Chinese tombs were often decorated with images of tortoises. Long ago, there was a strange belief in China that all tortoises were female and to reproduce they mated with snakes.

In Japan, the tortoise was regarded as one of the goddesses of love. In Bangkok there is a Buddhist temple in which turtles are kept in special pools. They are fed by the worshippers who believe that these animals are the temporary homes of human souls.

One tribe in New Guinea believes the thigh bone of a river turtle has magic properties and uses it to heal wounds and kill pain and they are prepared to pay a high price for one of these precious bones.

Turtles must have also been held in great esteem in Mongolia. A 2 m long stone monument of a turtle was found in the ruined city of Karakorum.

Turtles differ a great deal in both shape and size and some have a very strange appearance. The matamata turtle (*Chelus fimbriatus*), native to northern Brazil and Guyana, has a curious neck and head which may be longer than the whole shell. There are three rough ridges on its shell which run from front to back, and rough lobes of skin on its neck, so the animal looks like a pile of debris. Its extraordinary head is flat and has a flexible and extended nose. Unlike other turtles, this creature does not withdraw its head when molested, but curls its long neck to one side under the rim of the shell.

The matamata turtle feeds on fish and catches them by opening its mouth and sucking the fish and some water into its throat. It shuts its mouth and the prey is swallowed whole.

The world's largest tortoise is the Aldabra giant tortoise (*Geochelone gigantea*) found in the islands of Aldabra, Seychelles and Mauritius. The largest specimen weighed nearly 300 kg.

The most famous, however, is the Galapagos giant tortoise (*Geochelone elephantopus*), found on the Galapagos Islands. This huge tortoise can measure up to 1.2 m in length and weigh over 200 kg. They must have been much larger when Darwin visited the islands, since he reported that it took eight men to lift one giant tortoise from the ground. These huge tortoises provided large amounts of meat and, since they could be kept alive for many months without food, they became a superb source of fresh meat for the early sailors. One of these huge tortoises is said to have lived without food for six years. It is estimated that in the past no less than 10 million giant tortoises were taken from these islands by sailors.

The local inhabitants also used the giant tortoise as a source of water in an emergency. As Darwin noted: 'The inhabitants, when walking in the lower district, and overcome with thirst, often take advantage of this circumstance, and drink the contents of the bladder in full. In one I saw killed the fluid was quite limpid, and had only a very slightly bitter taste'.

Giant tortoises are among the longest-living animals and some are estimated to have lived up to 150 years. They are far older than dinosaurs and have remained relatively unchanged for some 150 million years.

The hierarchy among the Galapagos giant tortoises is based on the height to which each animal can extend its head. Their numbers have declined considerably and they are now under strict protection.

The largest turtles are found in the sea. The largest is the leatherback turtle (*Dermochelys coriacea*), and on average is 2 m long and weighs about 500 kg. The largest specimen was nearly 3 m long, nose to tail, and weighted about 960 kg. Interestingly, the males of this species never leave the sea water, and only the females come ashore to lay their eggs. In order to avoid predators, they leave the sea at night and lay their eggs in deep holes which they have to dig. One turtle may lay up to 100 eggs the size of ping-pong balls, cover them with sand and return to the sea. When the young hatch, they must find their way out through the sand and then rush to the sea. It is a crucial moment in their lives, since during their journey many perish attacked by various predators purposefully waiting on the beach.

Some turtle shells are incredibly strong. For example, the carapace of the North American box turtle (*Terrapene* species) can support a weight 200 times its own. A man of such strength could support the weight of two adult elephants.

Some turtles, such as the alligator snapping turtle (*Macroclemys temminckii*), found in rivers and lagoons from Canada to northwestern South America are hated by fishermen. These turtles use a 'bait' to catch fish. They have a peculiar outgrowth on their tongue which resembles a red worm. This 'bait' is like a living worm and is constantly moving to lure fish. Attracted by this peculiar bait, fish try to catch the 'worm', but are immediately caught in the clever turtle's mouth. The American fishermen are not at all happy with these turtles because they substantially reduce their own catches of larger fish.

Some turtles have been used by men for a rather unusual purpose. The common snapping turtle (*Chelydra serpentina*) has occasionally been used to detect a human corpse in a lake or river. This turtle is a carrion eater so it is particularly sensitive to the smell of rotting flesh and can detect a human body from a considerable distance.

The common snapping turtle has reportedly caused injuries to people who have tried to touch it in the water. It is said to be able to cut off a finger if molested.

Some sea and river turtles are known for their great migrations to lay their eggs. The green turtle (*Chelonia mydas*), which feeds off the

Brazilian coast, is known to migrate to the beaches of Ascension Island to nest, covering a distance of more than 2,000 km.

Natural Goalpost

The ostrich *(Struthio camelus)* is not only the largest of all living birds, but also the heaviest. It can be up to 2.7 m tall and can weigh up to 150 kg. The eye of this giant bird is as large as the entire body of the bee hummingbird. This remarkable bird lays the largest eggs in the avian world. One egg can be some 20 cm long and can weigh up to 1.8 kg. It would take nearly one hour to hard-boil such an egg and one ostrich egg would be sufficient to make an omelette for twelve people.

As well as the eggs being a source of food, the very hard shells have been utilised for various purposes. Bushmen use the empty shells for water containers, which are very efficient as they can be buried in the sand to keep the water cool. The Bushmen particularly use such containers when they make raids on neighbouring tribes to steal cattle. Before a raid, they bury several water-filled ostrich egg shells in places only known to themselves. They use this water when escaping after a raid and their enemies have to give up chasing them because they have no water to drink in the desert.

Ostrich eggs were popular containers before glass was discovered. Richly decorated egg-cups made of ostrich eggs were a proud possession of ancient Greek and Egyptian nobles. Beautiful, engraved ostrich egg cups have been found in graves dating from about 3000 BC.

African natives used to hunt ostriches extensively, using poisoned arrows. In order to approach the alert birds they would disguise themselves by wearing ostrich feathers. The hunter held up one hand to imitate the ostrich's head and neck and moved in an ostrich-like fashion to fool the bird.

Roast ostrich was a famous dish in the past. One Roman emperor once organised a great hunting expedition and ordered his men to bring him as many ostriches as they could find. Six hundred birds were killed, the story says, and their brains were served during a banquet to impress his guests.

This bird's beautiful feathers attracted people in ancient times even more than its meat. Since the ostrich feather is remarkably symmetrical, it was believed in Egypt that they represented the goddess of justice and truth and that they were used to weigh the virtues of souls after death.

African warriors used ostrich feathers to decorate their heads and shields, and the feathers also became popular decorations in Europe. In the nineteenth century, every elegant lady would adorn her hat with ostrich feathers. Such was the demand for these feathers that in 1909 alone, twenty tonnes of ostrich feathers, valued at half a million dollars, were imported to Europe from Africa.

Although ostriches cannot fly, they can run, for brief periods, at an astonishing speed of over 70 kmh. The ostrich is therefore the fastest two-legged animal on earth. It can outpace a galloping horse and escape predators. This gigantic bird seems to be devoid of fear and will even charge a moving train, with usually fatal consequences for the bird. It can be dangerous to man if cornered and a kick to the head can kill a man. Even a man on horseback can be overpowered by a high-kicking ostrich.

This dangerous animal can be tamed and trained to perform various tasks. It can be trained to pull a small chariot and ostriches are used, especially in South Africa, for racing where a light man or boy rides in each trotting cart. But probably the most peculiar use of tamed ostriches was found in South Africa, where a sportsman used them in a football match. He simply placed two ostriches at each end of the field, where they stood for hours as natural goalposts. Ostriches have also been used by farmers in some regions to chase off other birds attacking their valuable crops. Some farmers still use trained ostriches to herd their sheep.

The world's largest birds have many interesting habits. Contrary to a common belief, ostriches do not bury their heads in the sand to avoid being attacked. They only squat with their necks on the ground to become harder to see by a potential predator or egg thief. The male keeps a harem of three to four females and he keeps them in line. When he calls them from a distance, he makes a noise that can be likened to the roar of a lion. Strangely, this powerful boss shares in the incubation of the eggs with his females. The male, however, seems to be ashamed of his female-like function, as he only sits at night. Since many eggs are laid in the same nest by the females, some eggs near the edge could be stolen by a predator. Interestingly, each female seems to care especially for her own eggs, which she easily recognises, and always tries to push them to the inside of the heap, to make them less vulnerable to predators and better incubated.

Ostriches are particularly attracted to shiny objects and they can swallow coins, pieces of glass, rings and even small watches. There was even a report from South Africa that a valuable diamond was discovered in an ostrich's crop. Ostriches are known to swallow stones and sand and a baby bird starts swallowing small stones as soon as it hatches. It is said that these objects help their digestion.

These birds feed on vegetable matter and they relish fallen, ripe fruits. Sometimes birds consume a lot of overripe and fermented fruit and become drunk and as a result behave like a typical drunkard.

Ostriches are still a source of valuable feathers and meat and, in many regions, special ostrich farms have been established. There is an amusing story related to such a farm in Australia. When one of the first ostrich farms was established in Australia these birds were kept free in a field and, when the owners wanted to harvest the feathers, they discovered that they could not catch the birds even on horseback. In the Flinders Ranges in South Australia, one can see ostriches living free, as if on the African savannah. They are the descendants of the ostriches that once escaped from a local farm.

Nature's Skyscraper

The most impressive animal found in the East African savannas is no doubt the giraffe (*Giraffa camelopardalis*), the tallest land animal in the world. With its enormously elongated neck it stands some 5 m tall. It is such a huge animal that an adult man can stand upright between its front legs.

Giraffes have caused a number of problems in East Africa because of their enormous necks. A 1000 km long telegraph line system was established in Kenya, but just a few days after completion the lines were found broken in many places. Saboteurs were suspected of causing this damage and wardens were ordered to shoot them on sight. It was later found that tall giraffes were to blame. The animals were breaking the telegraph wires and even posts while crossing the roads and, to prevent further damage, all the wires had to be elevated to 9 m above the ground.

The world's tallest animals are very inquisitive and as a result, have also caused many problems. In 1908 a goods train travelling at night had to stop for several hours because of two giraffes, one of which one had poked its nose across the engine and broken the engine's lanterns. Galloping giraffes once even derailed a train when they collided with a running locomotive at night.

The giraffe is such a peculiar animal that the early writers claimed that the Creator must have run out of ideas if he had to combine the features of a camel and a hyena. An Arab scholar wrote in 1022 that the father of the giraffe was a leopard, because of its beautifully patterned coat, and the mother was a camel mare, because of the giraffe's long legs and camel-like pace. Small wonder that early naturalists called the giraffe the 'camelopard'.

The giraffe played an important role in the beliefs of the ancient Egyptians. Their god of evil, called Set, was always depicted as a man with the head of a giraffe. Such however, was the value of a live giraffe that when the Egyptians defeated the Nubians in the seventeenth century BC, they not only demanded human slaves, but also required one giraffe.

These strange animals were also admired in ancient times in Rome. They once had a giraffe at a sacrificial ritual to the god of Sun and Moon. When the horses saw this peculiar animal, the story says, they were so terrified that they disrupted the whole ceremony.

The giraffe is so different from other animals that when it was brought to the Imperial Zoological Garden in Peking in 1414, people watched the long-necked beast with great excitement. It was believed to be the mythical Confucian beast, Kilin. At one time, the giraffe was considered by the Chinese to represent perfect virtue and harmony and was regarded as the emblem of a perfect government. The ancient Arabs were less impressed with the beauty of giraffes, but to dream of this animal had special significance. It could mean a financial loss, bad news or an unfaithful wife. Surprisingly, this animal remained almost unknown in Europe until the end of the eighteenth century. Knowledge of the existence of this animal was based on ancient Greek writings and hearsay stories and it was considered to be a mythical or legendary beast.

In many regions of Africa, the ownership of live giraffes was a great privilege. Those who possessed a great number of giraffes were considered to be the most wealthy people. In the sultanates of Chad, a chief's power was related to the number of giraffes' tails he owned.

Despite its remarkable beauty, the giraffe was usually hunted by the natives in Africa. Only one tribe in Kenya, known as the Kamasya, conceived the image of their god Moa as a hornless giraffe.

Once a giraffe has been killed, all its parts can be used. Its meat is edible and buckets and drum coverings are made from its skin. The hide is also used for making charms against lions. The most valuable part of the giraffe however is its tail. Tails are used as amulets for good luck and for making bracelets, which are especially popular among the Masai women. Unfortunately, demand for hair bracelets for tourists was so strong that poachers killed them just for their tails. Tribes in Sudan and Ethiopia have traditionally hunted giraffes and this has led to their almost total extermination in some regions.

Despite the enormous length of its neck, the giraffe has the usual seven vertebrae like most other mammals, but each vertebra is about 40 cm long. Its long neck enables the animal to reach twigs on the tops of many trees which are usually beyond the reach of other browsing animals. Giraffes relish the leaves and twigs of acacia trees, which are covered with sharp thorns, but because of their thick lips and tough tongues, giraffes can eat the 'thorny' food without harm. The elastic tongue of the giraffe is, in fact, an important tool. Measuring some 45 cm, the tongue can be curled around objects and used like a hand. The tip of the giraffe's tongue is so fine that it can even be poked into a keyhole.

Because of their height, giraffes have the greatest range of vision among mammals and are quickly able to spot an enemy from a great distance. Although a giraffe can kill a lion with a powerful kick it always prefers to escape from a potential predator than to fight.

Despite her enormous height, the giraffe remains standing when giving birth. The baby has to drop from a height of some 1.5 m, but this does not cause it any harm. The new-born baby is quite strong and it can usually walk within an hour of birth.

Giraffes usually browse in groups, and it is a real pleasure to observe these graceful animals, especially in the Nairobi National Park, where they occur in great numbers. We had a chance to observe the giraffe's strange behaviour there one day. We saw a male approach a female and collect some of her urine, evidently 'testing' its hormone content by lifting his head and curling the lip. This was a typical 'check-up' to determine whether a female is in oestrus. If the 'test' is positive, the bull follows her until she stops and allows him to mount her. The male's penis is about 75 cm.

Giraffes are still very common in some parts of Africa, especially in Kenya where they can be seen freely walking along or across roads.

Drivers always give way to them. Imagine our amazement while driving near Nairobi in Kenya when a large giraffe blocked the whole road, standing in the middle and looking at us with her 'wise' eyes. We had to wait for about 10 minutes until, with dignity and without haste this tower-like animal moved gracefully away.

Index